Transport and Energy Processes

AA008952

Held at the 2023 AIChE Annual Meeting

Orlando, Florida, USA
5-10 November 2023

ISBN: 978-1-7138-9297-7

Printed from e-media with permission by:

Curran Associates, Inc.
57 Morehouse Lane
Red Hook, NY 12571

Some format issues inherent in the e-media version may also appear in this print version.

Printed with permission by Curran Associates, Inc. (2024)

For permission requests, please contact AIChE
at the address below.

AIChE
120 Wall Street, FL 23
New York, NY 10005-4020

Phone: (800) 242-4363
Fax: (203) 775-5177

www.aiche.org

Additional copies of this publication are available from:

Curran Associates, Inc.
57 Morehouse Lane
Red Hook, NY 12571 USA
Phone: 845-758-0400
Fax: 845-758-2633
Email: curran@proceedings.com
Web: www.proceedings.com

TABLE OF CONTENTS

POSTER SESSION: TRANSPORT AND ENERGY PROCESSES DIVISION

THERMAL ENERGY STORAGE

TRANSPORT IN ADVANCED FUEL CELL AND WATER ELECTROLYZER TECHNOLOGIES

ADVANCED ELECTROCHEMICAL ENERGY STORAGE TECHNOLOGIES II

TRANSPORT AND ENERGY PROCESSES AT ELECTROCHEMICAL INTERFACES I

TRANSPORT AND ENERGY PROCESSES AT ELECTROCHEMICAL INTERFACES II

146a Online Electrochemical Mass Spectrometry for the Investigation of Electrolyte Degradation in High Voltage Li-Ion Pouch Cells

Jane Edgington, Chi Cheung Su, Meinan He

Transportation is a key sector of the US economy while accounting for 27% of total US greenhouse gas emissions during 2020. The ongoing wide scale adoption of electric vehicles (EV) enables electrification of the transportation sector and the opportunity for significant reduction in greenhouse gas emissions. High voltage (>4.2V) battery development is critical as car manufacturers strive to develop high power, extended range EVs. One key issue surrounding high voltage battery performance is electrolyte stability. While some initial electrolyte degradation is desirable during formation to develop the SEI, overactive electrolyte degradation throughout extended cycling can lead to poor capacity retention and coulombic efficiency (CE), as well as excessive gas evolution and pouch pressurization. Electrolyte degradation typically results from oxidation and reduction reactions occurring at the interface between electrolyte and anode or cathode, drive by voltage extremes. To explore the high voltage stability of electrolyte and battery chemistries, we leverage online electrochemical mass spectrometry (OLEMS) to characterize online gas evolution in batteries. OLEMS is a powerful tool that we have used to identify gaseous and volatile electrolyte degradation products throughout battery cycling, as well as quantify gaseous species production rates as a function of time and voltage.

While OLEMS analysis of battery systems has predominantly employed coin cell-type cells, we have developed an OLEMS system that continually samples 2.5 Ah pouch cells using Ar carrier gas. By characterizing gas evolution in pouch cell samples, we have designed our DEMS system and experiments with an applications-based approach. We leverage our system to characterize gas evolution behavior during formation and initial C/3 cycling for ethylene carbonate/ethyl methyl carbonate (EC/EMC) and fluoroethylene carbonate/dimethyl carbonate (FEC/DMC) electrolyte systems under standard and elevated upper cutoff voltages of 4.2 and 4.5 V. We find that both electrolyte solvent and upper cutoff voltage condition greatly impact gas product species, production profiles during cycling, and gas quantity. Gas evolution of oxidation products, such as CO2 and CO, is exacerbated by the elevated upper voltage cutoff position, while reduction products are only minimally affected. Overall, we find the EC/EMC electrolyte is more stable against gas evolution at both upper and lower upper cutoff voltage conditions. Our reported work serves to inform future electrolyte development for high voltage Li-ion batteries, to demonstrate OLEMS as a powerful tool for gas evolution analysis, and to offer a fundamental understanding of gas evolution behavior for applications-based battery pouch cells.

146b Materials Informatics for Designing Optimal Electrolytes for Lithium-Sulfur Batteries

Nav Nidhi Rajput, Rasha Atwi

A major breakthrough in battery materials is required to meet the ever-increasing proliferation of portable electronic devices, electric vehicles and their variants, as well as the need for incorporating renewable energy resources into the main energy supply.[1] In this context, lithium-sulfur (Li-S) batteries attract attention owing to their very high energy density (2,600 Wh kg-1) and specific capacity (1,675 mAh g-1) and significantly lower weight and cost, compared to lithium-ion batteries (LIBs).[2] Fully packaged, it is expected that future Li-S batteries can operate at close to 500 Wh kg-1, which is more than twice the energy density of LIBs (200 Wh kg-1). The problem of realizing the expected high energy density is defined by several issues including the dissolution of Li-Polysulfide (PS) species into the electrolyte, insulating properties of sulfur and Li-PS species, and volume change at the cathode.[3] Overcoming these challenges requires a fundamental understanding of the interplay between events occurring over wide spatial and temporal scales, and accurate prediction of electrode and electrolyte properties to obtain design metrics for new improved materials.

In this talk, I will first discuss the details of a high-throughput multi-scale computational infrastructure developed by our group called **MISPR** (Materials Informatics for Structure-Property-Relationship).[4-6] MISPR seamlessly integrates density functional theory (DFT) calculations with classical molecular dynamics (MD) simulations and generates high-fidelity databases of computational properties and includes several fully automated workflows to compute electronic, thermodynamic, structural, and dynamical properties of electrolyte solutions.

I will then discuss the usage of MISPR to design optimal electrolytes for Li-S batteries by altering the atomistic interactions between the electrolyte components through high-throughput screening of potential co-solvent molecules. This approach guides and accelerates our rational selection of co-solvents that enable optimal compromise between the solubility of PS species and the transport properties of the electrolyte through automated DFT calculations. We use the selected candidates in detailed MD studies to comprehend the relationship between the structure of the co-solvent and the electrolyte properties. The approach allows for creating a database of well-characterized materials to be used in machine learning-based methods as well as for testing computationally identified structures in experiments. We recently published the first publicly available database, **ComBat** (Computational Database for Li-S Batteries), which includes ~2000 properties for solvents spanning 16 different chemical classes. This work provides crucial information to alleviate the dissolution of PS species during cycling while maintaining high ionic conductivity and low viscosity.[7, 8]

References:

1. Larcher, D.; Tarascon, J.-M., Towards greener and more sustainable batteries for electrical energy storage. *Nature chemistry* **2015,** *7* (1), 19-29.
2. Manthiram, A.; Yu, X.; Wang, S., Lithium battery chemistries enabled by solid-state electrolytes. *Nature Reviews Materials* **2017,** *2* (4), 1-16.
3. Manthiram, A.; Fu, Y.; Su, Y.-S., Challenges and prospects of lithium–sulfur batteries. *Acc. Chem. Res.* **2012,** *46* (5), 1125-1134.
4. Atwi, R.; Bliss, M.; Makeev, M.; Rajput, N. N., MISPR: an open-source package for high-throughput multiscale molecular simulations. *Scientific Reports* **2022,** *12* (1), 15760.
5. https://github.com/molmd/mispr.
6. Atwi, R.; Chen, Y.; Han, K. S.; Mueller, K. T.; Murugesan, V.; Rajput, N. N., An automated framework for high-throughput predictions of NMR chemical shifts within liquid solutions. *Nature Computational Science* **2022,** *2* (2), 112-122.
7. Atwi, R. a. R., Nav Nidhi, Guiding Maps of Solvents for Lithium-Sulfur Batteries via a

Computational Data-Driven Approach. *Patterns* http://dx.doi.org/10.2139/ssrn.4324048 **2023**.
8. https://github.com/rashatwi/combat.

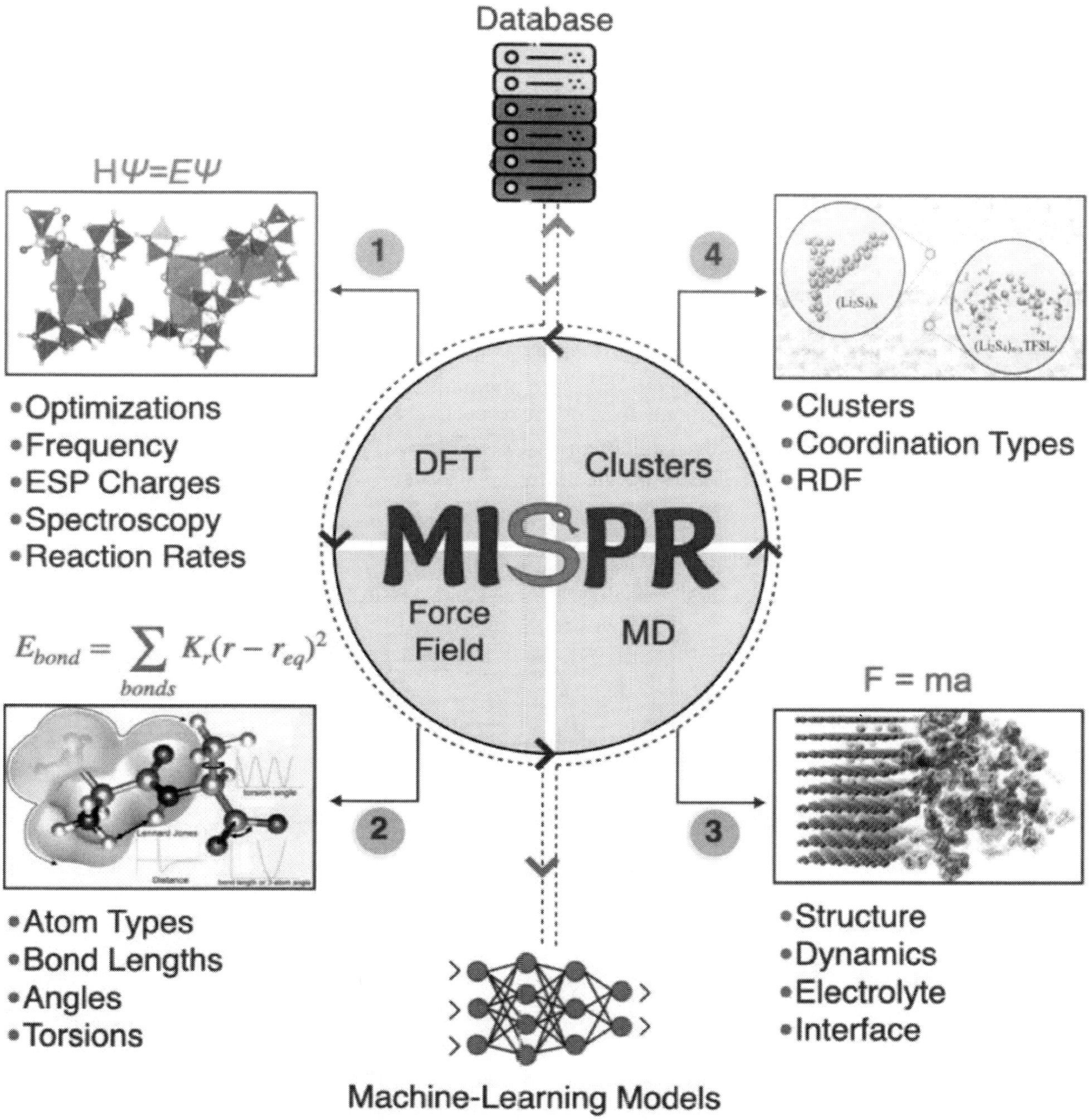

146d Eutectic Electrolytes for Secondary Al-CO_2 Batteries

Chris Fetrow, Cameron Carugati, Madalin Alderete, Shuya Wei

Metal-CO_2 batteries have emerged as a promising strategy for enhancing energy storage technology while capturing/concentrating carbon dioxide. We previously demonstrated an Al-CO_2 battery utilizing a homogeneous iodine-based redox mediator to enable the reversible discharge and charge of the battery with an ultra-low overpotential of 0.05V (Fetrow *et al., ACS Appl. Mater. Interfaces* **2023**, 15, 10, 12908–12914). The imidazole-based electrolyte employed earlier is common for aluminum stripping/plating but is expensive, corrosive, and hygroscopic. In this study, we investigate several alternative electrolytes. These include $AlCl_3$-based melts previously explored for aluminum-ion batteries, as well as novel AlI_3-based eutectic melts. As shown in **Figure 1**, both urea- and acetamide-$AlCl_3$ melts are compatible with the aluminum iodide homogeneous catalyst. We characterized electrolyte performance electrochemically and investigated the mechanisms enabling ionic liquid formation and performance using Raman and NMR spectroscopies. These findings facilitate the development of a practical Al-CO_2 battery.

Figure 1: Cyclic voltammetry and galvanostatic discharge studies of alternative ionic liquids for Al-CO_2 batteries. A, B) Cyclic voltammogram of coin cells with aluminum and stainless steel electrodes using 1.3:1 $AlCl_3$:acetamide and 1.3:1 $AlCl_3$:urea electrolytes, respectively. Dashed lines indicate cells without added 0.05M AlI_3. C) Galvanostatic discharge of an Al-CO_2 battery with aluminum and nanostructured carbon cathodes using a 1.3:1 $AlCl_3$:acetamide electrolyte containing 0.05M AlI_3. Discharged at 20 mA/g_{carbon} to a capacity of 620 mAh/g_{carbon}.

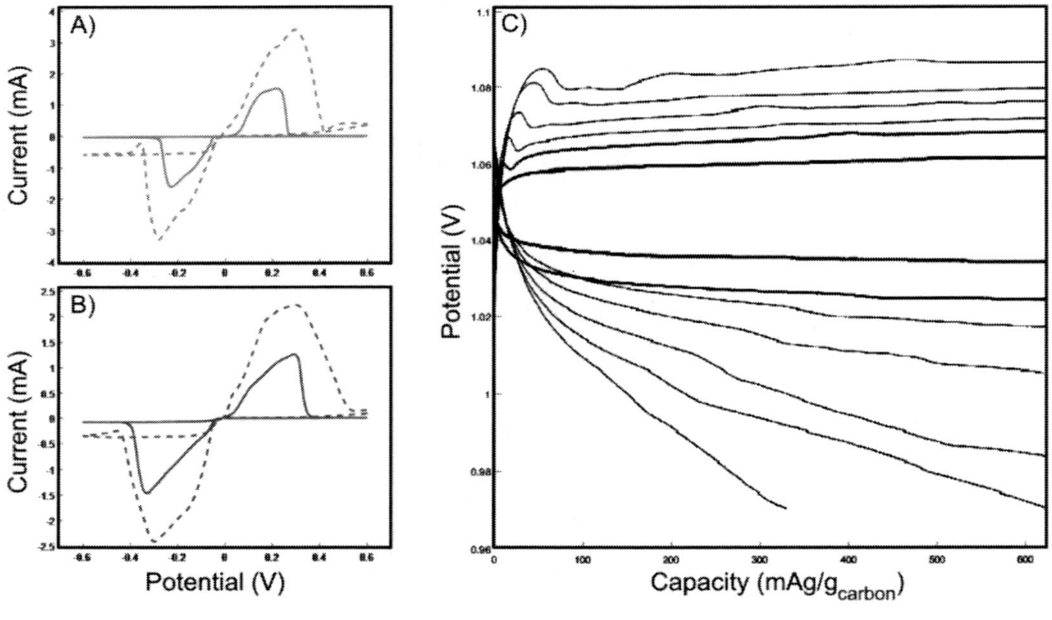

224i Investigations of Lithium Plating on Thick Graphite Electrodes in Fast Charging Lithium-Ion Batteries

Guanyi Wang, Aleksandar Mijailovic, Mei Luo, Jie Xiong, Bingyao Zhou, Kevin Mathew, Emmanuel Kornyo, Wenquan Lu, Brian Sheldon, Qingliu Wu

With the rapid growth of the electric vehicle (EV) market, fast charging capability is becoming one of the key features of lithium-ion batteries (LIBs). However, charging traditional LIBs with thick electrodes at high current densities generally adversely impact the electrochemical performance and cause potential safety issues due to the formation of lithium dendrites. The onset of Li dendrite formation is associated with the polarization of cells mainly originated from the retard mass transport across the electrodes. This study investigated the onset of Li plating in the graphite electrodes by using electrochemical analysis. After discharging (lithiating) the graphite electrode to different states of charge (SOC) at various rates, the open-circuit voltage (OCV) during the relaxing period was monitored and the analysis of differential open-circuit (dOCV) was applied to manifest the onset of Li plating. The onset of Li plating was further verified through the analysis of coulombic efficiency (CE) and post-mortem analysis on the cycled electrodes using scanning electron microscopy (SEM). The same approaches were used to study the onset of Li plating in the graphite electrodes with different thicknesses. Based on analysis from these tests, the correlation of Li plating onset, electrode thickness and lithiation rate was established. Furthermore, the experimental observations on the Li plating onset agree well with theoretical predictions obtained with a Newman pseudo-2D model. Therefore, the achievements obtained here will be of both scientific and practical significance in developing LIBs with high energy and power densities.

146f Amending Porous Electrode Theory through Recasting of Shrinking Core Models

Zeeshan Chaudhry, Alexander Couzis, Ilona Kretzschmar

Porous electrodes are often employed in the fabrication of electrochemical systems since many of their manufacturing processes entail using methods developed by the powder science and technology field. Therefore, models of porous electrodes have been logically constructed from pre-existing powder science and technology models. However, key features regarding the interfacial behavior of these systems have yet to be fully integrated. For example, in electrode systems that participate in plating and stripping processes, each particle is subject to a change in size. Subsequently, this alters the particle size distribution that comprises the porous electrode and the void volume occupied by the electrolyte in the pores, where mass transfer is relevant. At the particle scale, this gives rise to a moving boundary problem that affects the performance of porous electrodes. For instance, if a porous electrode is initially consolidated and begins to participate in an electro-stripping process, the particles will begin to shrink, as shown in Figure 1 using zinc as an example.

Consequently, the overall resistivity of the porous electrode increases due to particles losing contact with one another and a decrease in the available pathways for current to enter and exit through the current collector. Furthermore, parasitic reactions may lead to passivation upon reversing the polarity across the cell, which otherwise would ideally induce electroplating. Some particles may never be re-incorporated into a connected network of particles. Thus, effectively rendering those particles electrochemically inactive.

Therefore, this presentation uses concepts from interfacial phenomena, reaction engineering, and transport phenomena to formulate descriptions of porous electrodes. This is done by considering a contemporary example, the metallic zinc anode, which is prevalent in many commercially available battery systems. This work first considers one particle in a porous electrode, through which parallels to the shrinking core model from reaction engineering are made and eventually used to revise traditional porous electrode models.

Zinc-anode electrochemistry is a multifaced problem worth examining. This complexity arises from the electrochemical cell housing the anode comprising a solid-liquid interface under galvanostatic conditions (i.e., constant current or non-equilibrium conditions), where the anode-electrolyte interface is of interest. Therefore, both the metallic Zn-anode and the alkaline electrolyte contribute to the behavior of the interface and the associated problems on its surface. On the electrolytic side, elementary processes such as adsorption, desorption, and surface reactions (electrochemical and thermochemical) can be used to model the anode interface. The mass transfer dynamics on the electrolytic side supplies the reactants necessary to initiate processes at the interface in the form of these elementary processes. On the electrode side, the current distribution governs the surface potential and energetics at the interface that instigates the desired redox reactions from which power is sourced from the cell. As a result, this makes the study of a metallic anode a complex and dynamic problem. The intent is to understand the rate limitations of the anode operation as a function of its environmental and operating conditions. This mechanistic understanding will be attained by formulating a theoretical and simulation model that are informed by appropriate experiments and validated by experiments as well. In this manner, the processes at the electrode-electrolyte interface can be interrogated to provide a dynamic model, which will provide a bridge for scaling the system for industrial applications.

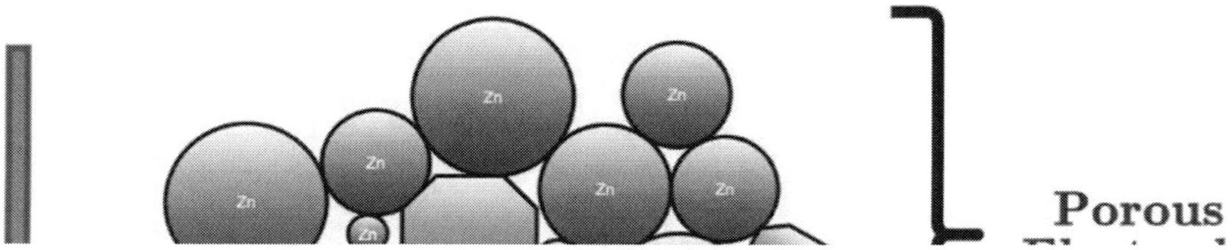

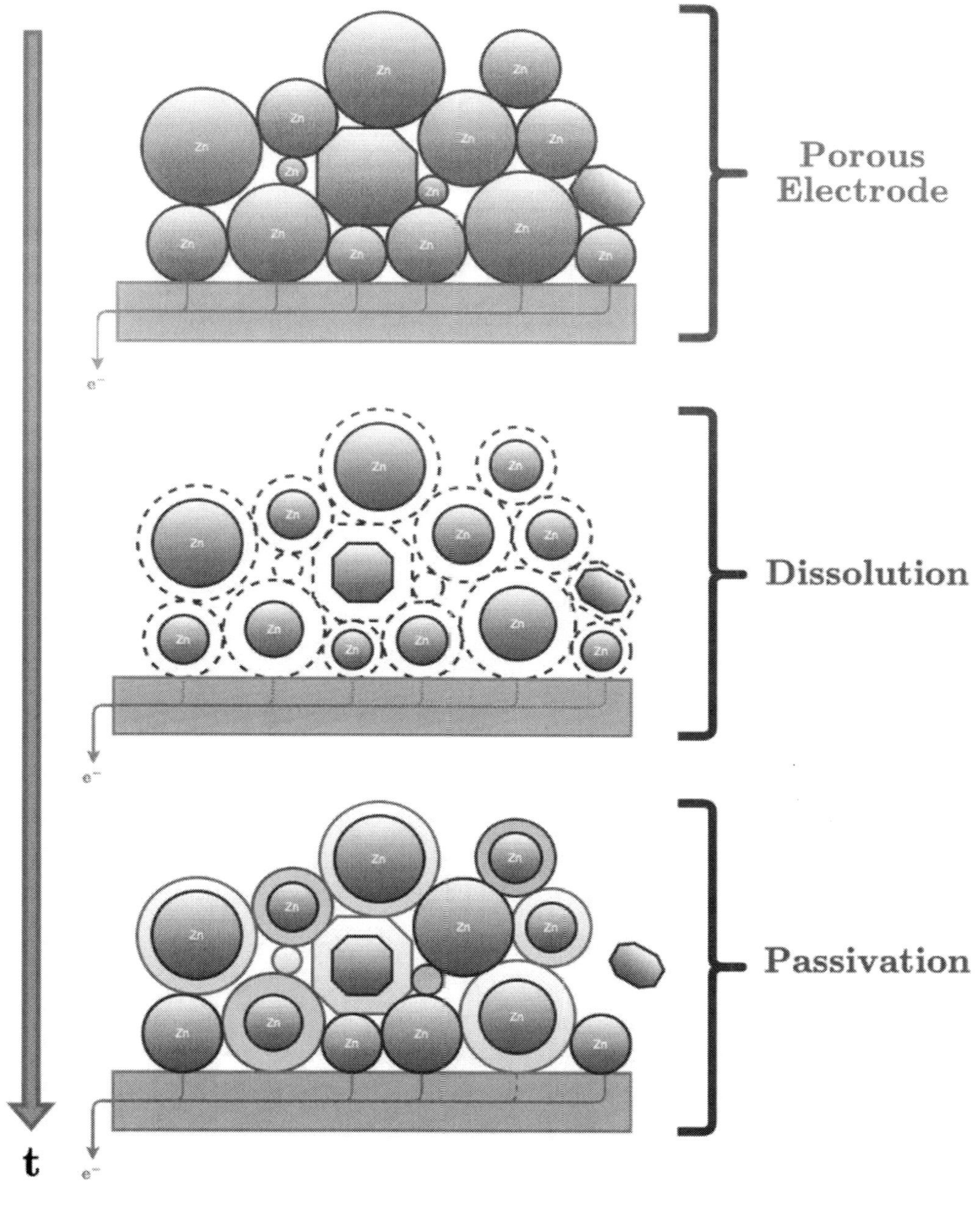

Figure 1: Schematic of moving boundaries and evolution of resistivity or conductivity of electrode material over time.

146g Advanced Fluorescence Microscope and Smart Cloud Algorithm in Real-Time Quantitative Processing and Analysis of Electroconvection

Duhan Zhang, Lynden A. Archer

In an electrochemical cell at high current densities, electroconvection plays a critical role in determining the morphology of electrodeposited metals. The resultant dendrite formation leads to risky battery performance failure. Over the past decades, though intensive theoretical analysis and experimental effort have been done, the detailed structure and dynamics of electroconvection remain unknown, because experimental observations lacked the resolution to expose the fundamental, hierarchical spatial structure of these flows, while the data processing algorithms were not fully established for large sample size. To address this shortcoming, we built an advanced optical electrochemical cell that fits in situ imaging with a super-resolution fluorescence microscope. By observing and recording real-time motions of electrolytes, electroconvection flows have been interrogated at unprecedented, high temporal and spatial resolution. As a complement to the visualization studies, a cloud-computing analysis algorithm was developed by combining a higher resolution Particle Image Velocimetry algorithm and a machine learning model to enable velocity distribution data over the entire optical field of view at nanoscale or microscale spatial resolution. Consequently, detailed velocity maps are obtained across the optical electrochemical cell, allowing us to quantitatively measure and analyze the initiation and evolution of the hierarchical microstructures inside the electroconvection in a unidirectional electric field. With these powerful tools, we further studied the effect of polymer viscoelasticity on electroconvection and electrodeposition. The addition of an ultrahigh molar mass polyethylene oxide to the electrolytes transforms the materials to a viscoelastic fluid state that changes the electroconvection dependence on time and voltage and induces a smooth electrodeposition on the metal anode by unique polymer rheological properties. Especially, the relaxation of long polymer chains in electrolyte provides a promising manner to prevent dendrite growth. We further analyzed these observations using direct numerical simulations for Newtonian and viscoelastic fluid models, and the quantitative analysis of experimental results indicates good consistency with simulation predictions.

316a Unraveling the Role of Topology in the Thermal Transport Limits of Metal-Organic Frameworks: A High-Throughput Study.

Meiirbek Islamov, Hasan Babaei, Jeffrey R. Long, Alan J. H. McGaughey, Diego Gomez Gualdron, Christopher Wilmer

Metal-Organic Frameworks (MOFs) are a unique class of crystalline materials that possess high porosity and surface area, making them promising candidates for various applications, including gas storage, separation, and catalysis. However, their practical utilization is often hindered by the challenge of managing the exothermic heat generated during gas adsorption. Despite their potential, the thermal transport properties of MOFs have received relatively little attention, leading to a lack of fundamental knowledge about the structure-thermal conductivity relationships required for designing MOFs with specific thermal properties. To address this gap, we performed a comprehensive high-throughput screening of hypothetical MOFs using classical molecular dynamics simulation and the Green-Kubo method. We screened 10,194 hypothetical MOFs that are generated using the ToBaCCo-3.0 code, considering several important structural and compositional features, such as density, pore size, surface area, void fraction, node-linker mass mismatch, and metal node connectivity. Our analysis revealed that small pores, high density, small node-linker mass mismatch, and four-connected metal clusters are associated with high thermal conductivity. Additionally, we identified six hypothetical MOF structures with exceptional thermal conductivity and 36 structures with ultra-low thermal conductivity. Interestingly, the six MOFs with ultrahigh thermal conductivity share a common feature of four-fold coordinated metal nodes that connect the organic linkers perpendicularly. This observation suggests that the topology of MOFs plays a crucial role in determining their thermal conductivity limits. Overall, our study provides new insights into the design principles of MOFs with tailored thermal properties and highlights the importance of considering the structural characteristics in designing high-performance MOFs for various applications.

316b Impact of Interfacial Properties on Carrier Transport and Photoelectrode Performance in Metal-Insulator-Semiconductor Structures

Alex J. King, Adam Z. Weber, Alexis T. Bell

Photoelectrochemical (PEC) cells are of significant interest for their potential to convert solar energy into chemical bonds, thereby enabling the storage of solar energy in forms that can be transported and used on demand. A critical challenge that PEC systems face is the design and fabrication of stable photoelectrodes that can effectively absorb light and catalyze the reaction. This challenge has led many researchers to investigate metal-insulator-semiconductor (MIS) structures as photoelectrodes because the metal catalyst has lower kinetic overpotentials than the bare semiconductor surface at the same current density and the ultrathin insulator layer helps to protect the semiconductor from corrosion by the electrolyte. The ultrathin insulator also serves as a barrier for which charge carriers must tunnel through, and previous modeling work has shown how the insulator thickness can be tuned in order to enable selective tunneling of the desired charge carrier. This selective carrier tunneling helps to minimize carrier recombination in the metal and, thereby, enable higher current densities. These models, however, fail to explain how selective carrier tunneling fundamentally impacts the MIS photovoltage and subsequent fuel-formation rates as well as the effects of interfacial non-idealities. Continuum modeling is uniquely suited to elucidate the effects of carrier tunneling and non-idealities on MIS performance and to identify interfacial properties beyond the insulator thickness that can be leveraged for improved solar-to-fuel (STC) conversion.

This talk will present recent continuum modeling efforts to simulate an MIS photoelectrode used for PEC water splitting. Our simulations have identified how properties of the insulator modulate carrier tunneling rates and how these rates impact the MIS photovoltage. Specifically, tuning the insulator properties affects the probability of tunneling, which alters the carrier concentrations at the semiconductor surface. Increasing the carrier concentrations at the semiconductor surface increases the photovoltage (*i.e.*, qausi-Fermi level splitting) by reducing the magnitude of the electric field and by increasing the quasi-Fermi level of the minority carrier. Our model has revealed optimal properties of the insulator that promote carrier buildup at the semiconductor surface and, thereby, enable high MIS performance. These predicted optimal properties of the insulator are a large bandgap (resulting in a high band offset with the semiconductor), a high dielectric constant, and a thickness of ~ 1.2 nm. Insulating materials that exhibit a large band offset with the semiconductor are predicted to lead to high MIS performances by creating an appreciable interfacial tunneling resistance that promotes carrier buildup at the semiconductor surface, and the moderate insulator thickness allows for unimpeded tunneling of the desired carrier. Increasing the thickness of these insulators beyond their optimal value results in hindered carrier tunneling and a very large concentration of carriers at the semiconductor surface. This extremely large carrier concentration causes a significant number of charges to be trapped at defect sites at the semiconductor-insulator interface, which results in a substantial drop in potential across the insulator that reduces the MIS photovoltage. Understanding how other interfacial properties can be engineered in order to mitigate the potential drop across the insulator is critically important for designing MIS photoelectrodes that can reach its theoretical maximum performance. The results of this study provide great insight into the mechanisms for which interfacial transport affects photoelectrode performance and provides optimal interfacial properties that tune carrier transport. Such an understanding is vital for the development of next-generation MIS photoelectrodes that enable highly efficient STC energy conversion.

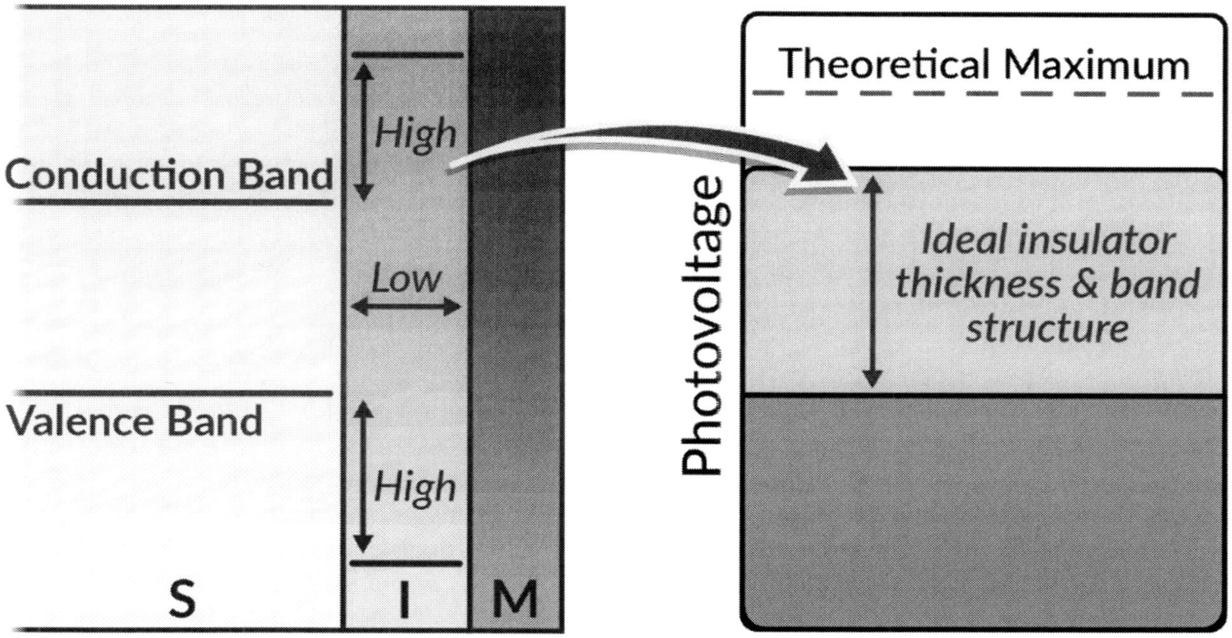
Conduction Band
Valence Band
High
Low
High
S
I
M
Theoretical Maximum
Photovoltage
Ideal insulator thickness & band structure

316c Alternative Approaches to Minimizing Loss Pathways in InP-Based Air-Bridge Cells

Bosun Roy-Layinde, Jihun Lim, Tobias Burger, Stephen Forrest, Andrej Lenert

The use of selectively reflective photovoltaic cells in thermophotovoltaic energy systems enables the recovery of otherwise unusable low energy (out-of-band) photons, thereby increasing the power conversion efficiency of the technology. We recently developed a thin-film cell architecture with a nanoscale air cavity buried beneath the light-absorbing InGaAs layer [1]. This cell exhibits ~98.5% out-of-band (OOB) reflectance, enabling conversion efficiencies exceeding 32% under illumination by ~1450 K SiC emitter. To investigate the performance of this device at different emitter temperatures (or power densities), we characterize the cell under various conditions characteristic of high thermal load [2]. We also quantify the loss pathways (including but not limited to Ohmic losses, thermalization losses, non-radiative recombination and OOB losses) associated with varying emitter temperatures. In this work, we propose two distinct cell configurations that further leverage the unique air-bridge architecture to overcome several challenges associated with operation of TPV cells at varying temperatures. First among these configurations is a double airbridge tandem cell, which leverages the top and bottom Au grid lines to optically and electronically connect the sub cells, thereby eliminating the need for a conventional tunnel junction. This may help to mitigate some of the primary loss pathways, including Ohmic and thermalization losses. Finally, we describe an alternative approach to photon recuperation, in which a transparent cell allows transmission of unusable photons to a secondary emitter, thereby further eliminating the OOB loss characteristic of reflective cells [3]. Here, the use of the air-bridge architecture uniquely enables transfer of the thin-film epilayer to a transparent device substrate.

REFERENCE

[1] D. Fan, T. Burger, S. McSherry, B. Lee, A. Lenert, and S. R. Forrest, "Near-perfect photon utilization in an air-bridge thermophotovoltaic cell," *Nature*, vol. 586, no. 7828, pp. 237–241, 2020, doi: 10.1038/s41586-020-2717-7.

[2] B. Roy-Layinde *et al.*, "Sustaining efficiency at elevated power densities in InGaAs airbridge thermophotovoltaic cells," *Solar Energy Materials and Solar Cells*, vol. 236, p. 111523, 2022, doi: https://doi.org/10.1016/j.solmat.2021.111523.

[3] T. Burger, B. Roy-Layinde, R. Lentz, Z. J. Berquist, S. R. Forrest, and A. Lenert, "Semitransparent thermophotovoltaics for efficient utilization of moderate temperature thermal radiation," *Proceedings of the National Academy of Sciences*, vol. 119, no. 48, p. e2215977119, 2022, doi: 10.1073/pnas.2215977119.

316d Enhanced Cycling Performance of Iron Metal Batteries for Low-Cost Energy Storage

Jing Liu

Renewable energy sources, like solar and wind, can decarbonize our energy generation and help to address the climate change grand challenge. However, their intermittent nature makes integrating them into the grid challenging. Battery energy storage can buffer the mismatch between renewable electricity generation and grid demand, but its high cost is a hurdle to wide deployment. Fe is the second most abundant metal in the earth's crust and is the most-produced metal commodity, outperforming other metals like zinc and lithium. In addition, the Fe metal also has a high capacity, so the material-level energy storage cost is only $0.06/kWh, making it extremely promising for achieving the U.S. Department of Energy's cost target for large-scale grid energy storage

However, the cycle life of Fe metal batteries is limited by the low Coulombic efficiency of the Fe deposition/stripping reaction. Aqueous electrolytes based on Fe chloride or sulfate salts are commonly used in Fe metal batteries, but they suffer from low Coulombic efficiency (<91%) under mild operating conditions due to the undesirable hydrogen evolution reaction (HER).

To address this issue, a series of novel Fe electrolytes reinforced with Mg ions (FERMI) and Ca ions (FERCI) have been developed in this study. The FERMI and FERCI electrolytes exhibit significantly improved Coulombic efficiency, higher conductivity, and better deposition morphology compared to the baseline $FeCl_2$ electrolyte. Adding 4.5 M $MgCl_2$ or CaCl2 into the $FeCl_2$ electrolyte enhances the Fe deposition/stripping efficiency to 99.1%, significantly improving the cycling performance of Fe metal batteries in both half-cells and full-cells.

Experiments and computational studies reveal that the enhanced Fe deposition/stripping efficiency can be attributed to two reasons (Figure): (1) The presence of Mg^{2+}/Ca^{2+} leads to Fe deposition with a larger particle size and a smaller surface area, which tends to form less dead Fe during stripping. (2) Suppression of the HER through reduced free water, stabilization of the water structure toward reduction resistance, and modification of the Fe solvation structure.

These novel electrolytes not only enable a highly reversible Fe metal anode for low-cost energy storage technologies but also have the potential to address the HER side reaction problem in other aqueous electrochemical technologies, especially acidic systems, such as CO_2 reduction, NH_3 synthesis, and iron electrochemical making.

The findings presented in this study are significant for the development of low-cost and high-performance Fe metal batteries for grid-scale energy storage applications. The FERMI and FERCI electrolytes hold great potential for the commercialization of Fe metal batteries and have the potential to pave the way for more efficient and sustainable energy storage technologies.

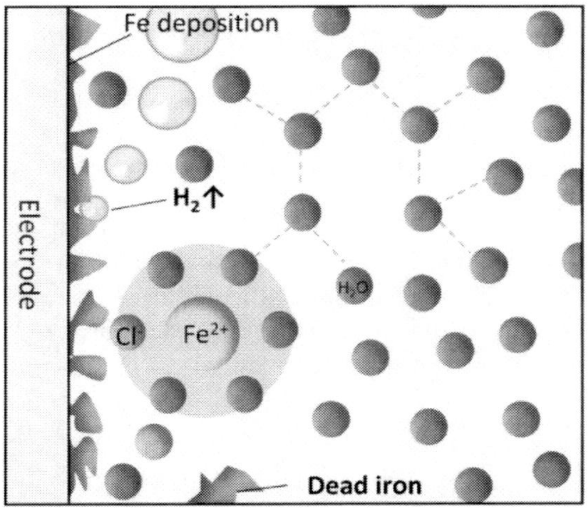
Fe deposition
Electrode
H_2↑
H_2O
Cl⁻
Fe^{2+}
Dead iron

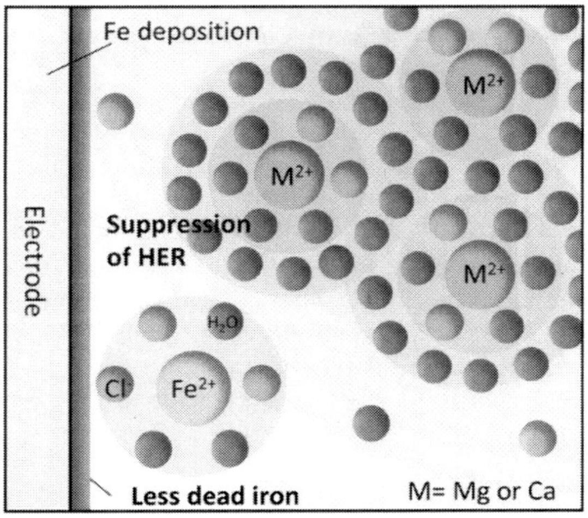
Fe deposition
Electrode
M^{2+}
M^{2+}
M^{2+}
Suppression
of HER
H_2O
Cl⁻
Fe^{2+}
Less dead iron
M= Mg or Ca

280a Modeling of Gas-Solid Phase Change in Cryogenic Carbon Capture

Jeffrey Young, Julia Hartig, Abhishek Venketeswaran, Michael McNelis, Phillip Buelow, Jonathan Stickel, Miles Abarr

Cryogenic carbon capture is a method to remove carbon dioxide from flue gas by cooling the gas mixture until the CO_2 desublimates as a solid. Simulating this process is necessary to predict the rate and location of solid CO_2 formation. There are three fundamental components that must be modeled to capture the desublimation dynamics: 1) the phase equilibrium, 2) the rate of phase change, and 3) the thermal effect of the phase change. These thermodynamic and transport processes are combined into a fundamental physics-based model for CO2 desublimation from flue gas. The phase-change model will be presented along with a comparison of the simulation results to experimental measurements. This model can enable better scalability of Carbon America's carbon capture technology.

280b Catalytic Regeneration of Aqueous Potassium Sarcosinate Solvent to Improve Energy Efficiency of CO_2 Direct Air Capture

Dhruba Jyoti Deka, Abishek Kasturi, Gyoung Gug Jang, Radu Custelcean, Costas Tsouris

Absorption of CO_2 in a chemical solvent is technologically a matured carbon capture method. However, cost-effective commercialization is yet to be realized due to the high energy requirement of solvent regeneration. The total regeneration energy (ΔH_{total}) is a combination of sensible heat (ΔH_{sen}) required to raise the solvent temperature, heat of vaporization (ΔH_{vap}) needed to vaporize a part of the solvent, and heat of CO_2-desorption (ΔH_{des}). Potassium sarcosinate (K-SAR) is an attractive direct air capture solvent but needs heating at temperatures in excess of 100°C for an extended period to release the absorbed CO_2. The current work aims to develop catalyst-aided solvent regeneration that can operate at a lower temperature (<100°C) to reduce the sensible and vaporization heats of K-SAR, as well as provide faster CO_2 desorption kinetics, thereby decreasing the overall energy requirement.

The experimental procedure involved heating 100-mL CO_2-loaded 3M K-SAR solvent on a heater-stirrer with or without catalysts. Various catalytic materials were used in these tests to evaluate their effectiveness: Fe_2O_3, TiO_2, SiO_2, and Fe-ZSM-5 (0.5g catalyst for 100 mL of solvent). The extent of solvent regeneration and its kinetics were determined by analyzing the CO_2 content in the evolved gas with a CO_2 sensor, as well as solvent samples via an acid-base titration. Combined thermogravimetric analysis (TGA-DSG) and FTIR experiments were also performed to quantify the overall regeneration energy and isolate the contributions of ΔH_{sen}, ΔH_{vap}, and ΔH_{des}.

Results obtained with Fe-ZSM5 catalysts showed that 30% better solvent regeneration, in terms of energy requirements, can be achieved in the presence of the catalyst than without it. These preliminary results clearly demonstrate that the regeneration energetics and kinetics of K-SAR solvent can be improved significantly by a catalyst. Such a concept of catalyst-assisted regeneration can also be extended to other common CO_2-capture solvents investigated for point-source capture, such as MEA.

280c Direct Air Capture of Carbon Dioxide Using Nickel Hydroxide Batteries in a Membrane Electrode Assembly

James Buchen

Direct air capture (DAC) is a growing field responding to the need to remove carbon dioxide from the atmosphere. Several technologies are being developed today attempting to meet this need. The technology furthest in development are temperature swing adsorption technologies which have recently moved out of the pilot plant scale.[1] This presentation looks to present an electrochemical device as alternative to these technologies.

The electrochemical device proposed is a pair of nickel hydroxide ($Ni(OH)_2$) battery electrodes which produce hydroxide (OH^-). The hydroxide reacts with carbon dioxide (CO_2) to create carbonate (CO_3^{2-}) and bicarbonate (HCO_3^-). These carbonates can be transported from the cathode across an anion exchange membrane to the anode where they are evolved back into CO_2 due to a pH reduction at the anode. This builds on work that optimized a Hydroxide Exchange Membrane Fuel Cell (HEMFC) for CO_2 capture, showing the efficacy of an electrochemically mediated pH swing carbon capture device.[2,3]

The benefit of the nickel hydroxide battery approach is the low energy cost associated with separation. Because a pair of nickel hydroxide batteries undergo the same electrochemical reaction, low potentials, and thus low energy cost is required. Experiments have shown device level energy requirements of less than 1 MWh $ton^{-1}CO_2$. Most of the energy required by the device produces the pH gradient used to capture and release CO_2.

The biggest hurdle for this electrochemical approach is increasing the flux of CO_2 to allow for more compact devices. Strategies for managing the transient charge and discharge behavior of the battery system will be discussed. Focus will be on improving flux to reduce the overall cost of the device.

References

1 Gambhir, A. & Tavoni, M. Direct Air Carbon Capture and Sequestration: How It Works and How It Could Contribute to Climate-Change Mitigation. *One Earth* **1**, 405-409 (2019). https://doi.org:10.1016/j.oneear.2019.11.006

2 Matz, S., Setzler, B. P., Weiss, C. M., Shi, L., Gottesfeld, S. & Yan, Y. Demonstration of Electrochemically-Driven CO2 Separation Using Hydroxide Exchange Membranes. *Journal of The Electrochemical Society* **168** (2021). https://doi.org:10.1149/1945-7111/abd5fe

3 Shi, L., Zhao, Y., Matz, S., Gottesfeld, S., Setzler, B. P. & Yan, Y. A shorted membrane electrochemical cell powered by hydrogen to remove CO2 from the air feed of hydroxide exchange membrane fuel cells. *Nature Energy* **7**, 238-247 (2022). https://doi.org:10.1038/s41560-021-00969-5

$$Cathode: \quad NiO(OH) + H_2O + e^- \rightarrow Ni(OH)_2 + OH^-$$

$$Membrane: \quad CO_2 + 2OH^- \rightarrow CO_3^{2-} + H_2O$$

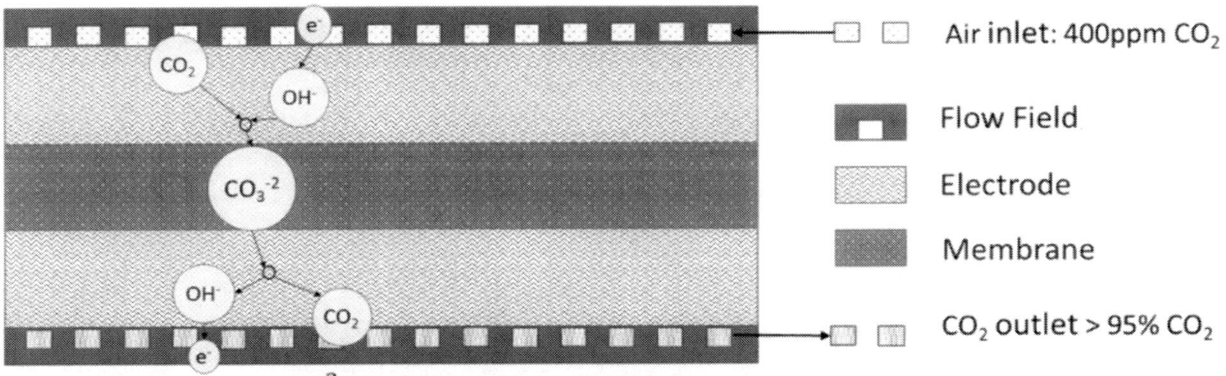

$$Ionomer: \quad CO_3^{2-} + H_2O \rightarrow CO_2 + 2OH^-$$

$$Anode: \quad Ni(OH)_2 + OH^- \rightarrow NiO(OH) + H_2O + e^-$$

280d Leveraging the Competitive Sorption Effects and Versatile Synthetic Strategy of Novel Microporous Poly(arylene ether) Membranes for Efficient CO_2 separations

Jing Ying Yeo, Sheng Guo, Duha Syar, Timothy Swager, Zachary Smith

Membrane-based gas separations are crucial for obtaining low-energy raw materials and reducing carbon emissions in the chemical industry as they offer an attractive non-thermal means to separate molecules. However, existing membrane technologies often suffer from intrinsic limitations of the membrane materials themselves. Microporous organic polymers (MOPs) are highly desirable for gas separation due to their combination of molecular sieving ability and solution-processability. In this work, we focus on a new class of MOPs within the family of poly(arylene ethers) (PAEs) for efficient CO_2 separations. The microporous structure of these PAEs, which are synthesized using a versatile strategy, is enhanced with CO_2-philic groups like nitrile and tertiary amine groups to improve the membrane interaction with CO_2 molecules, thereby enhancing CO_2 separation performance. The robust synthetic procedure and CO_2 competitive sorption effects of these polymers increase CO_2/CH4 selectivity by nearly 250% for mixed-gas tests relative to pure-gas tests. The work is extended to investigating the effects of highly plasticizing impurities like H_2S on competitive sorption under relevant conditions. Ternary $CO_2/CH_4/H_2S$ permeation studies reveal that H_2S enhances acid gas separation capability, permitting the simultaneous exclusion of H_2S and CO_2 in the mixture. This can reduce or even eliminate the need for pre-treatment steps in purification, thereby lowering the cost and complexity of the eventual separation processes. Besides, the branch-like architectures of these polymers provide stability for CO_2 separations that has never been observed for other membranes that were fabricated using purely linear MOP materials. This new family of PAEs provides significant and generalizable trends that can be leveraged to improve CO_2 separations for this unique class of materials. The methodology, implications, and future directions of applying this new class of PAEs to CO_2 separations will be outlined in this presentation, along with an in-depth literature survey on MOP permeation data for similar conditions.

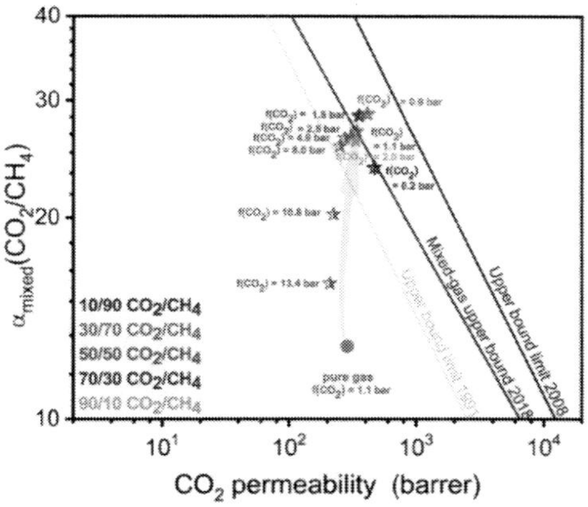

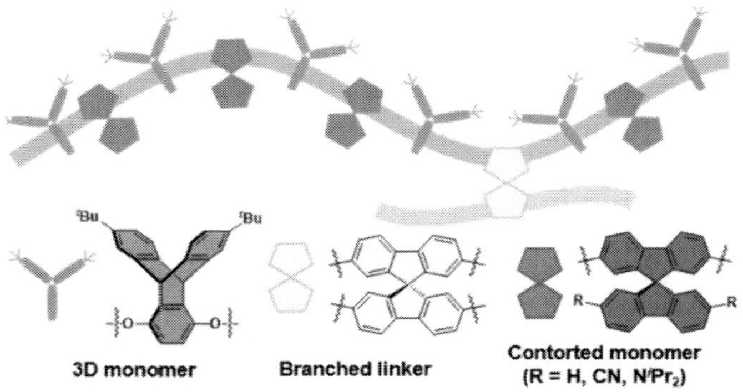

280e From Powder to Pellet: A Study of $La_{1-X}Ba_XFeO_3$ for Low-Temperature CO_2 to CO Conversion By a Reverse Water-Gas Shift Chemical Looping Process

Hanzhong Shi, Collin Schmidt, Jiawei Guo, Venkat R Bhethanabotla, John Kuhn

Reverse water-gas shift chemical looping (RWSG-CL) is a two-step process to convert carbon dioxide to carbon monoxide at a moderate temperature with high selectivity. In this research, perovskite oxides materials, $La_{1-x}Ba_xFeO_3$ (LBF, $x = 0.25 - 0.75$), were explored for enhancing CO yield while maintaining stability over multiple cycles. Then, the SiO_2-supported LBF (LBF/SiO_2) was synthesized to further enhance the conversion yield. LBF/SiO_2 pellets formed by tableting and extrusion methods were revealed the conversion properties. Density functional theory (DFT) was utilized to calculate the oxygen vacancy formation energies (E_{vac}), identifying the E_{vac} of $La_{0.5}Ba_{0.5}FeO_3$ (3.0 eV) is in the ideal range of 3.4 ± 0.5 eV. Meanwhile, the reaction mechanism was revealed by combined methods of DFT and diffuse reflectance infrared Fourier transform spectroscopy (DRFITS). For all samples, XRD confirmed cubic perovskite structure, and XRF analysis established the precise composition. The reaction temperature, approximately 550°C, was determined through temperature-programmed experiments. High material stability was proven by long-term RWGL-CL experiments combined with XRD. The CO_2 to CO yields were calculated in these experiments as 2.21, 2.41, and 2.35 mmol/g_{LBF} of 10 mm tableting, 6 mm tableting, and extrusion pellets, respectively. All pellets showed stable redox properties in 50 cycles of semi-batch reactor experiments, indicating that LBF/SiO_2 pellets are candidate materials for further scale-up evaluation.

280f Investigating the Effects of Ion Exchange Membrane Properties on the Performance of a Hydrogen-Powered Electrochemically-Driven CO_2 Separator

Marco Colin Martinez, Brian P. Setzler, Yushan Yan

The increasing need to reduce carbon dioxide emissions may be insufficient to meet the Paris Climate Agreement which demands new avenues for carbon capture. Particularly, direct air capture will become vital to meet these capture demands. Although traditional carbon capture technology has focused on adsorbent/absorption technology, they pose prohibitively large energy and capital costs. An alternative CO_2 capture approach has been demonstrated by exploiting the alkaline environment generated by anion exchange membrane (AEM) hydrogen fuel cells. Here, hydroxides are electrochemically generated at the cathode which chemically scrubs CO_2 out of the air, producing (bi)-carbonates that transport across the AEM and accumulate at the anode. There, the local pH environment lowers until CO_2 generation is thermodynamically favorable resulting in a concentrated CO_2 stream. From the stoichiometry of the proposed acid-base mechanism, the EDCS has a theoretical upper limit of 1 CO_2: e^- (mol captured/mol consumed), typified as 100% electron efficiency. Previous work on the electrochemically-driven CO_2 separator (EDCS) for DAC applications with 400 ppm CO_2 focused on understanding and improving the gas transport of CO_2 to the cathode. However, there is still a need to understand the effect of ion transport on performance. As seen in Figure 1, the intersection between the dashed and solid lines indicates an optimum for high electron efficiencies and CO_2 removal. At current densities near this optimum, the back-diffusion of carbonates is hypothesized to reduce the capture rate of CO_2 and negatively affect electron efficiency. The membrane resistance can be increased to improve performance as illustrated in Figure 1. Although membrane resistance increases with increasing membrane thickness, thicker membranes would result in higher capital costs. Thus an alternative approach is to decrease the ion exchange capacity of the membrane for a given membrane thickness. PAP-TP-85 has been previously employed for the EDCS, and it is known that at high temperatures (>80 oC), the ion exchange sites of the polymer are unstable in their hydroxide form but stable in their halide form. Therefore, by heating PAP-TP-85 at various hydroxide-to-chloride ratios, a method for selectively degrading ion exchange sites in hydroxide form has been developed. The membranes are characterized after degradation by employing potentiometric titration to determine IEC and EIS to measure membrane resistance. The impact of membrane resistance is examined in the EDCS on a range of IECs that have been prepared using the PAP-TP-85 membrane. Preliminary evidence shows improvement in electron efficiency at higher membrane resistance in the lower current range. This may suggest that diffusion plays a more important role at the lower currents when hydroxides are limiting in the CO_2 capture process. To meet the \$100/tCO_2 requirement, it is crucial to optimize energy efficiencies and CO_2 flux for lower operating and capital costs.

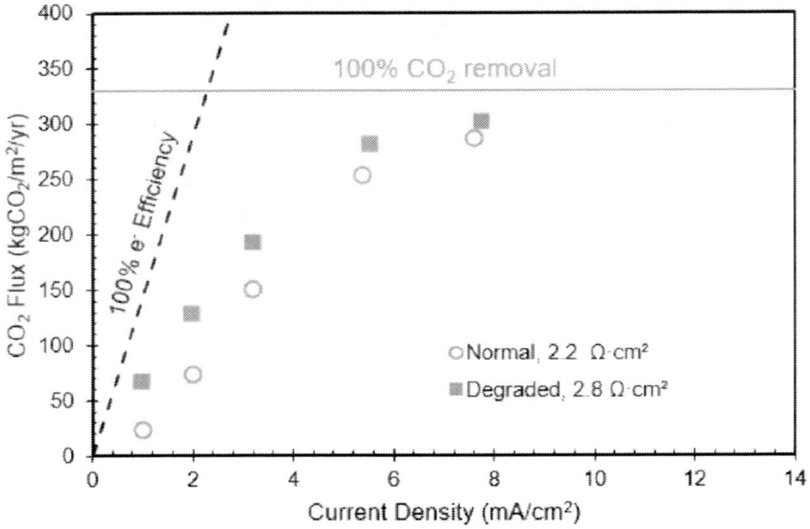

Figure 1: Preliminary data comparing the flux performance of a non-degraded and degraded PAP-TP-85 polymer membrane (50 ± 2 μm) performed with 2000 sccm air containing 400 ppm CO_2 at 40 °C (70%RH). The dashed black line represents the upper limit where all hydroxide produced at the cathode captures a CO_2 molecule. The horizontal green line indicates the maximum CO_2 flux possible based on the total CO_2 feed. The catalyst was 0.1 mg/cm² Pt for both the anode and cathode. A carbon-ionomer interlayer with 0.5 mg/cm² Vulcan XC-72 and PAP-TP-85 was added between the membrane and cathode catalyst layer. The ionomer was PAP-TP-85 with an ionomer-to-carbon weight ratio of 0.4 for the catalyst layers and the cathode interlayer.

280g Study of Vapor-Liquid Phase Equilibrium in High-Pressure CO_2-H_2O System Using Molecular Simulations

Atefeh Ahmadbeigi, Sumit Sharma

Carbon dioxide (CO_2) capture and sequestration (CCS) is a way of mitigating anthropogenic climate change. In CCS, the captured CO_2 is transported in supercritical or dense phase conditions from the source to the storage via pipelines. Corrosion of these pipelines, which occurs in the presence of free water and impurities, is a major concern. As CCS technologies achieve wider adoption in the near future, it has become imperative that safe operating conditions for these pipelines are established. We need to identify conditions that result in the precipitation of dissolved water from the CO_2 phase, thereby creating corrosive conditions. Many experimental studies have been done on the phase behavior of the water-CO_2 system. But most research has focused on low-pressure and temperature conditions. It is challenging to experimentally determine the solubility of water in dense phase CO_2. Furthermore, role of impurities and metallic interfaces on the nucleation of free water has remained unexplored.

In this work, I will discuss results obtained from fully atomistic simulations to study the phase behavior of water in dense phase CO_2 and will discuss the role of impurities.

2023 AIChE ANNUAL MEETING
November 5 — November 10, 2023
Hyatt Regency Orlando
Orlando, FL

Application of Doped Iron Oxides for the Thermochemical Conversion of CO_2 into Fuels

(Principal Author)
Rahul Bhosale
Department of Chemical Engineering,
University of Tennessee, Chattanooga, TN, 37421, USA

(Co-Authors)
Sayma Akhter
Department of Chemical Engineering,
Qatar University, Doha, Qatar

Prepared for Presentation at
American Institute of Chemical Engineers
2023 AIChE Annual Meeting
Orlando, FL, USA
November 5 – 10, 2023

Application of Doped Iron Oxides for the Thermochemical Conversion of CO_2 into Fuels

Rahul R. Bhosale*[,1], Sayma Akther[2]

[1]Department of Chemical Engineering, University of Tennessee, Chattanooga, TN, 37421, USA
[2]Department of Chemical Engineering, Qatar University, Doha, Qatar

Keywords: Doped Iron Oxide, Combustion Synthesis, CO_2 Conversion

Abstract

In this work, we synthesized doped iron oxides such as Co-doped iron oxides using the glycine-assisted solution combustion synthesis method. Powder X-ray Diffraction (PXRD) and Energy Dispersive Spectroscopy (EDS) are used to understand the phase composition of the synthesized doped iron oxide samples. Likewise, the Scanning Electron Microscope (SEM) is utilized to study the morphology of the derived doped iron oxides. The thermochemical reactor arrangement is also discussed.

1. Introduction

The excessive release of greenhouse gases such as CO_2 is one of the potential reasons for air pollution. Likewise, the incessant production and release of wastewater from industries is the prime reason for water pollution. Many efforts are underway to treat the abovementioned pollutants [1–14]. However, in addition to the treatment technologies, there is a pressing need to find solutions to avoid producing and releasing air and water pollutants, such as generating and applying alternative fuels instead of fossil fuels[15–31]. To be specific, one of the options to reduce the production and release of CO_2 into the atmosphere is to re-utilize the CO_2 to produce fuels[32–36]. It is possible to convert CO_2 into fuels in various ways, and one of the ways is to use the solar-assisted thermochemical CO_2 splitting cycle. As shown below, the CO_2 can be converted into fuel such as CO via metal oxide-based redox reactions. In the first step, the metal oxide is thermally reduced with the help of an inert atmosphere and solar energy input. The reduced metal oxide re-oxidizes in the second step via the CO_2-splitting non-solar step.

$$MO \rightarrow MO_{red} + \frac{\delta_{red}}{2} O_2 \qquad (1)$$

$$MO_{red} + (\delta_{red})CO_2 \rightarrow MO + (\delta_{red})CO \qquad (2)$$

Since the 1970s, various metal oxides have been utilized in solar thermochemical cycles[37–77]. Recently, it was observed that including suitable dopants in the base metal oxide crystal structure helps improve the solar thermochemical cycle's redox reactivity and overall efficiency. Hence, in this study, we have attempted to prepare doped iron oxide materials using a solution combustion approach to produce solar fuels.

Note: Do not add page numbers. Do not refer to page numbers when referencing different portions of the paper

2. Synthesis and Characterization Procedure

In this study, we have synthesized the doped iron oxide materials using the driven solution combustion synthesis method. At first, the precursors required, such as iron nitrate nonahydrate and other metal nitrates (used as dopants), were purchased from Sigma Aldrich, USA. Pre-treatment was not applied as the nitrates obtained had an extremely high purity. De-ionized water was used to dissolve the metal nitrates, and glycine was expended as the combustion fuel. The first step associated with the combustion synthesis was the molar calculation to determine the exact amounts of metal nitrates and fuel needed to attain a pre-determined doped iron oxide composition. Once the calculation was done, the weighted metal nitrates were dissolved in de-ionized water with the help of a continuous stirring mechanism. Glycine was added to the solution containing completely dissolved metal nitrates. This metal precursor + glycine solution was pre-heated to 150 to 200°C for complete water evaporation. Once the water from the solution was evaporated, a viscous gel formation was observed. This gel was further heated up to the auto-ignition temperature (300 to 400°C), and the solution combustion was initiated. Once the combustion was over, the powder obtained was removed from the breaker and crushed using a pestle and mortar. The crushed solid was annealed using a furnace up to 1000°C. The annealed powder obtained was further stored for the characterization. The powder was characterized using a powder X-ray Diffractometer (PXRD) and a Scanning Electron Microscope (SEM) to understand its phase composition and material morphology.

3. Results and Discussion

As explained in the paragraph above, the doped iron oxides were synthesized using a combustion approach. The entire combustion synthesis process is explained in Fig. 1. The crystal structure of iron oxide was doped with various metal cations from the transition metal section. The PXRD analysis helped confirm the doped iron oxide's phase composition and crystallite size. The exemplary PXRD peaks for the Co-doped iron oxide are shown in Fig. 2. The chemical compositions of the derived doped iron oxides were also confirmed by conducting the EDS analysis. SEM analysis was performed to scrutinize the morphology of the combustion-derived doped iron oxides. The SEM analysis confirmed the formation of nanoparticulated porous morphology. A sample SEM image of the Co-doped iron oxide is represented in Fig. 3.

In addition to synthesizing and characterizing the combustion-derived doped iron oxide materials, we have also developed a TGA-based thermochemical reactor set-up to test the redox reactivity of the derived samples. Fig. 4 shows the schematic of the TGA set-up installed in the laboratory.

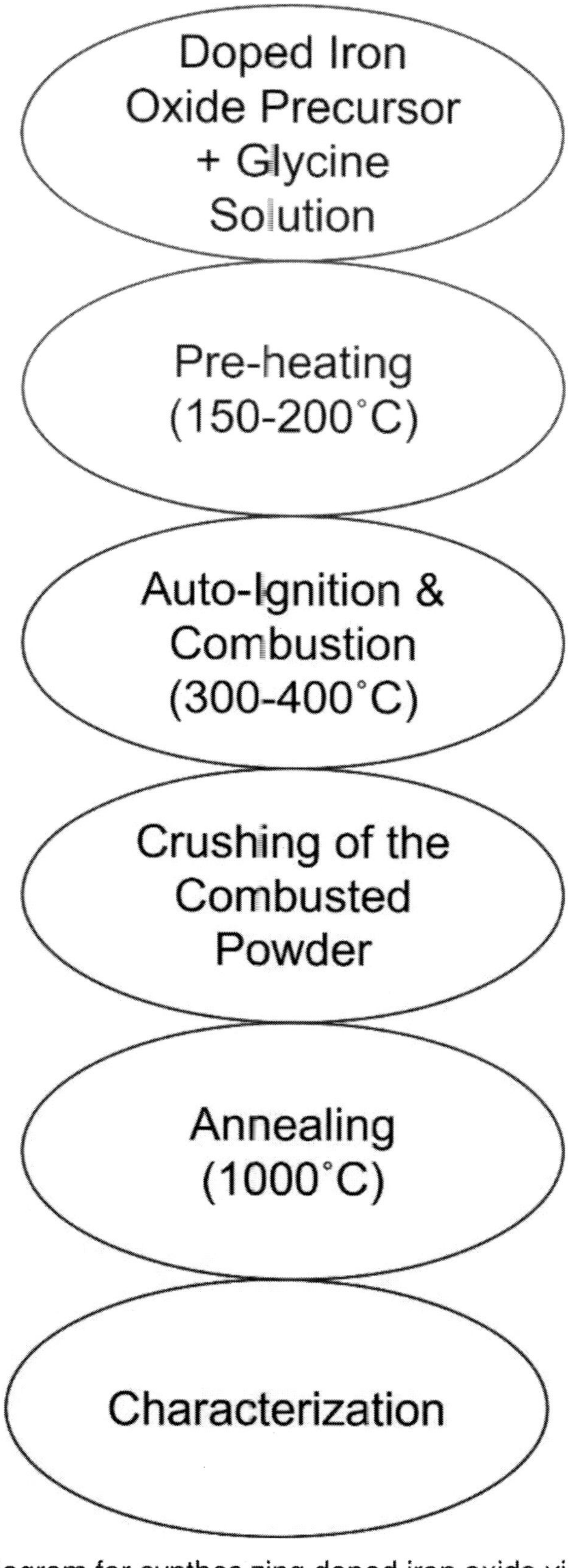

Fig. 1. Process flow diagram for synthes zing doped iron oxide via combustion method.

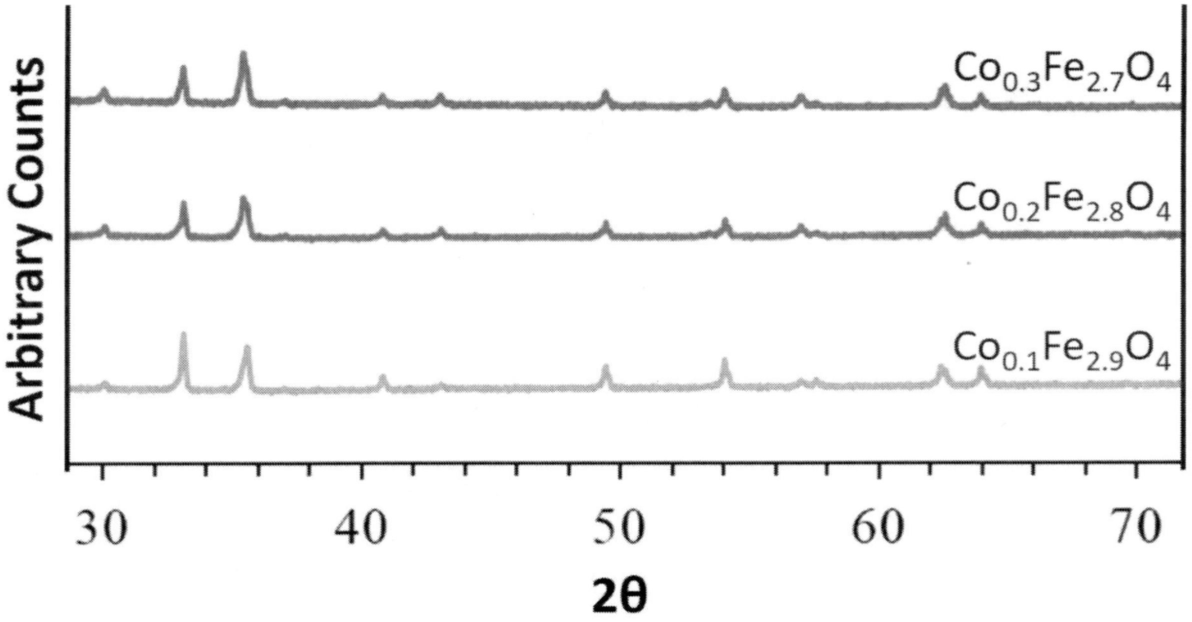

Fig. 2. PXRD peaks for Co-doped iron oxide.

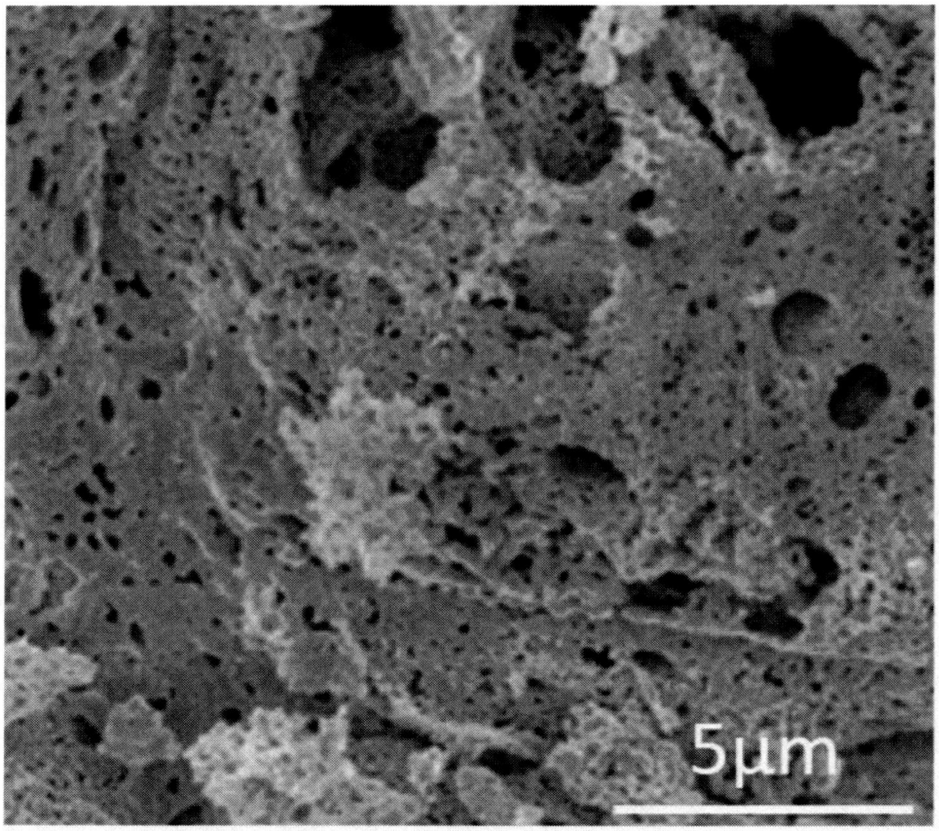

Fig. 3. SEM image of $Co_{0.2}Fe_{2.8}O_4$.

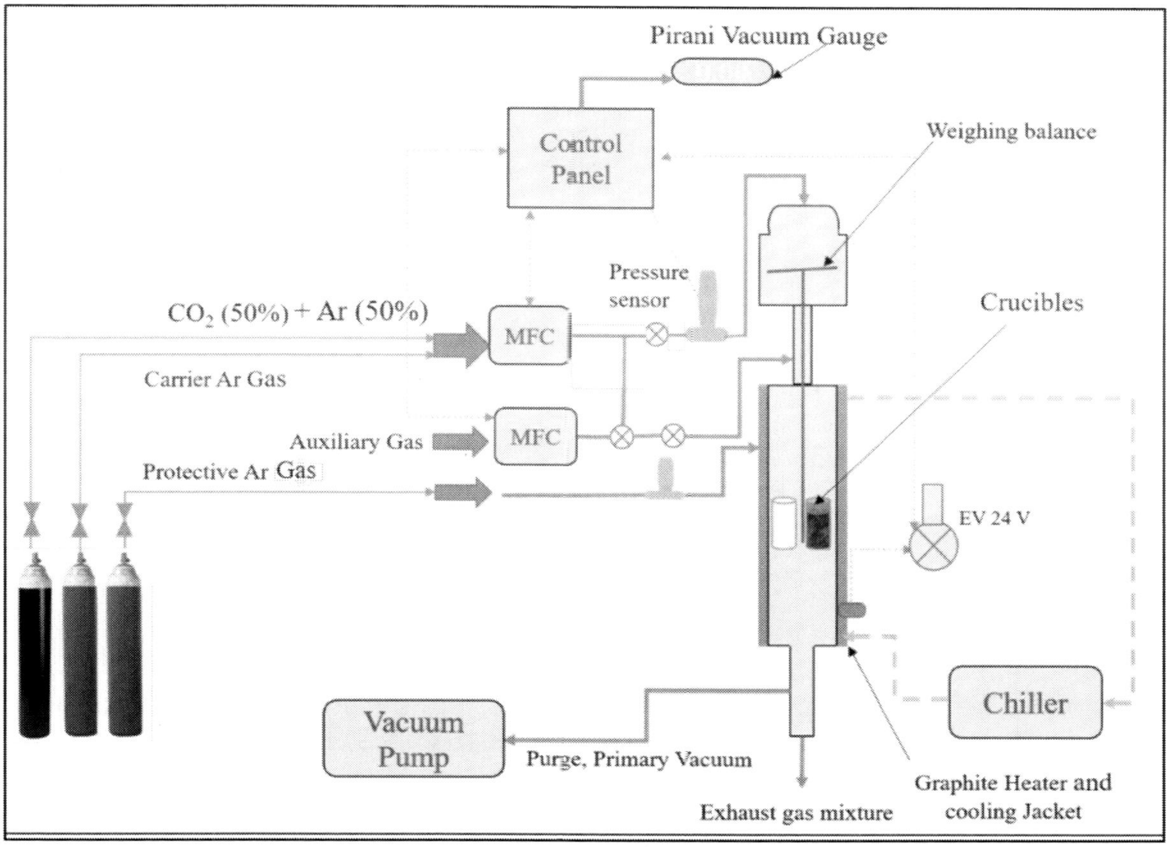

Fig. 4. The schematic of the TGA set-up installed in the laboratory.

4. Conclusions

Transition metal-doped iron oxide samples were synthesized using a solution combustion synthesis method. The PXRD analysis confirmed the formation of phase pure composition, and the SEM analysis indicated nanoparticulated porous morphology. A TGA-based thermochemical reactor was also developed and installed in the laboratory.

5. Acknowledgments

The authors would like to acknowledge the support provided by Qatar University International Research Collaboration Co-funds Grant No. IRCC-2021-002 and the Ruth S. Holmberg Grant for Faculty Excellence, University of Tennessee, Chattanooga, TN, USA.

References

[1] Bhosale RR, Kumar A, AlMomani F, Ghosh U, AlNouss A, Scheffe J, et al. CO_2 Capture Using Aqueous Potassium Carbonate Promoted by Ethylaminoethanol: A Kinetic Study. Ind Eng Chem Res 2016;55:5238–46. https://doi.org/10.1021/acs.iecr.5b04398.

[2] Vaidya PD, Kenig EY. CO2-Alkanolamine Reaction Kinetics: A Review of Recent Studies. Chem Eng Technol 2007;30:1467–74. https://doi.org/10.1002/ceat.200700268.
[3] Gulied M, Al Momani F, Khraisheh M, Bhosale R, AlNouss A. Influence of draw solution type and properties on the performance of forward osmosis process: Energy consumption and sustainable water reuse. Chemosphere 2019;233:234–44. https://doi.org/10.1016/j.chemosphere.2019.05.241.
[4] Zhang QH, Yang WN, Ngo HH, Guo WS, Jin PK, Dzakpasu M, et al. Current status of urban wastewater treatment plants in China. Environ Int 2016;92–93:11–22. https://doi.org/10.1016/j.envint.2016.03.024.
[5] Mohamad S, Fares A, Judd S, Bhosale R, Kumar A, Gosh U, et al. Advanced wastewater treatment using microalgae: effect of temperature on removal of nutrients and organic carbon. IOP Conf Ser Earth Environ Sci 2017;67:012032. https://doi.org/10.1088/1755-1315/67/1/012032.
[6] Ahmed SN, Haider W. Heterogeneous photocatalysis and its potential applications in water and wastewater treatment: a review. Nanotechnology 2018;29:342001. https://doi.org/10.1088/1361-6528/aac6ea.
[7] Bhosale RR, Mahajani V V. Kinetics of thermal degradation of renewably prepared amines useful for flue gas treatment. Journal of Renewable and Sustainable Energy 2013;5. https://doi.org/10.1063/1.4831960.
[8] Moreira D, Pires JCM. Atmospheric CO2 capture by algae: Negative carbon dioxide emission path. Bioresour Technol 2016;215:371–9. https://doi.org/10.1016/j.biortech.2016.03.060.
[9] Bhosale RR, Mahajani V V. Kinetics of Absorption of Carbon Dioxide in Aqueous Solution of Ethylaminoethanol Modified with N-methyl-2-pyrolidone. Sep Sci Technol 2013;48:2324–37. https://doi.org/10.1080/01496395.2013.805228.
[10] Stolaroff JK, Ye C, Nguyen DT, Oakdale J, Knipe JM, Baker SE. CO2 Absorption Kinetics of Micro-encapsulated Ionic Liquids. Energy Procedia 2017;114:860–5. https://doi.org/10.1016/j.egypro.2017.03.1228.
[11] Almomani F, Bhosale R, Kumar A, Khraisheh M. Potential use of solar photocatalytic oxidation in removing emerging pharmaceuticals from wastewater: A pilot plant study. Solar Energy 2018;172:128–40. https://doi.org/10.1016/j.solener.2018.07.041.
[12] Toth AJ. Modelling and Optimisation of Multi-Stage Flash Distillation and Reverse Osmosis for Desalination of Saline Process Wastewater Sources. Membranes (Basel) 2020;10:265. https://doi.org/10.3390/membranes10100265.
[13] Almomani F, Bhosale RR. Bio-sorption of toxic metals from industrial wastewater by algae strains Spirulina platensis and Chlorella vulgaris: Application of isotherm, kinetic models and process optimization. Science of The Total Environment 2021;755:142654. https://doi.org/10.1016/j.scitotenv.2020.142654.
[14] Crini G, Lichtfouse E. Advantages and disadvantages of techniques used for wastewater treatment. Environ Chem Lett 2019;17:145–55. https://doi.org/10.1007/s10311-018-0785-9.
[15] Almomani F, Bohsale RR. Optimizing nutrient removal of moving bed biofilm reactor process using response surface methodology. Bioresour Technol 2020;305:123059. https://doi.org/10.1016/j.biortech.2020.123059.

[16] Tungal R, Shende R V. Hydrothermal liquefaction of pinewood (Pinus ponderosa) for H2, biocrude and bio-oil generation. Appl Energy 2014;134:401–12. https://doi.org/10.1016/j.apenergy.2014.07.060.

[17] Pugazhendhi A, Nagappan S, Bhosale RR, Tsai P-C, Natarajan S, Devendran S, et al. Various potential techniques to reduce the water footprint of microalgal biomass production for biofuel—A review. Science of The Total Environment 2020;749:142218. https://doi.org/10.1016/j.scitotenv.2020.142218.

[18] Rodionova MV, Poudyal RS, Tiwari I, Voloshin RA, Zharmukhamedov SK, Nam HG, et al. Biofuel production: Challenges and opportunities. Int J Hydrogen Energy 2017;42:8450–61. https://doi.org/10.1016/j.ijhydene.2016.11.125.

[19] Shurair M, Almomani F, Bhosale R, Khraisheh M, Qiblawey H. Harvesting of intact microalgae in single and sequential conditioning steps by chemical and biological based – flocculants: Effect on harvesting efficiency, water recovery and algal cell morphology. Bioresour Technol 2019;281:250–9. https://doi.org/10.1016/j.biortech.2019.02.103.

[20] Voloshin RA, Rodionova M V., Zharmukhamedov SK, Nejat Veziroglu T, Allakhverdiev SI. Review: Biofuel production from plant and algal biomass. Int J Hydrogen Energy 2016;41:17257–73. https://doi.org/10.1016/j.ijhydene.2016.07.084.

[21] Rajesh Banu J, Yukesh Kannah R, Kavitha S, Ashikvivek A, Bhosale RR, Kumar G. Cost effective biomethanation via surfactant coupled ultrasonic liquefaction of mixed microalgal biomass harvested from open raceway pond. Bioresour Technol 2020;304:123021. https://doi.org/10.1016/j.biortech.2020.123021.

[22] Matin MA, Kumar A, Bhosale RR, Saleh Saad MAH, Almomani FA, Al-Marri MJ. PdZn nanoparticle electrocatalysts synthesized by solution combustion for methanol oxidation reaction in an alkaline medium. RSC Adv 2017;7:42709–17. https://doi.org/10.1039/C7RA07013F.

[23] Almomani F, Shawaqfah M, Bhosale RR, Kumar A, Khraisheh MAM. Intermediate ozonation to enhance biogas production in batch and continuous systems using animal dung and agricultural waste. Int Biodeterior Biodegradation 2017;119:176–87. https://doi.org/10.1016/j.ibiod.2016.11.008.

[24] Abanades S, Abbaspour H, Ahmadi A, Das B, Ehyaei MA, Esmaeilion F, et al. A critical review of biogas production and usage with legislations framework across the globe. International Journal of Environmental Science and Technology 2022;19:3377–400. https://doi.org/10.1007/s13762-021-03301-6.

[25] Almomani F, Bhosale RR, Khraisheh MAM, Shawaqfah M. Enhancement of biogas production from agricultural wastes via pre-treatment with advanced oxidation processes. Fuel 2019;253:964–74. https://doi.org/10.1016/j.fuel.2019.05.057.

[26] Kabeyi MJB, Olanrewaju OA. Biogas Production and Applications in the Sustainable Energy Transition. Journal of Energy 2022;2022:1–43. https://doi.org/10.1155/2022/8750221.

[27] Rodriguez JA, Liu P, Stacchiola DJ, Senanayake SD, White MG, Chen JG. Hydrogenation of CO_2 to Methanol: Importance of Metal–Oxide and Metal–Carbide Interfaces in the Activation of CO_2. ACS Catal 2015;5:6696–706. https://doi.org/10.1021/acscatal.5b01755.

[28] Ashok A, Kumar A, Bhosale R, Saleh Saad MA, AlMomani F, Tarlochan F. Study of ethanol dehydrogenation reaction mechanism for hydrogen production on combustion synthesized cobalt catalyst. Int J Hydrogen Energy 2017;42:23464–73. https://doi.org/10.1016/j.ijhydene.2017.01.175.

[29] GO K, SON S, KIM S. Reaction kinetics of reduction and oxidation of metal oxides for hydrogen production. Int J Hydrogen Energy 2008;33:5986–95. https://doi.org/10.1016/j.ijhydene.2008.05.039.

[30] Kumar A, Ashok A, Bhosale RR, Saleh MAH, Almomani FA, Al-Marri M, et al. In situ DRIFTS Studies on Cu, Ni and CuNi catalysts for Ethanol Decomposition Reaction. Catal Letters 2016;146:778–87. https://doi.org/10.1007/s10562-016-1706-9.

[31] Li J, Kazakov A, Dryer FL. Experimental and Numerical Studies of Ethanol Decomposition Reactions. J Phys Chem A 2004;108:7671–80. https://doi.org/10.1021/jp0480302.

[32] Almomani F, Bhosale R, Khraisheh M, Kumar A, Tawalbeh M. Photocatalytic conversion of CO2 and H2O to useful fuels by nanostructured composite catalysis. Appl Surf Sci 2019;483:363–72. https://doi.org/10.1016/j.apsusc.2019.03.304.

[33] Dhabarde N, Selvaraj J, Yuda A, Kumar A, Subramanian VR. Review of photocatalytic and photo-electrocatalytic reduction of CO2 on carbon supported films. Int J Hydrogen Energy 2022;47:30908–36. https://doi.org/10.1016/j.ijhydene.2022.02.124.

[34] Ashok A, Kumar A, Bhosale RR, Almomani F, Saleh Saad MAH, Suslov S, et al. Influence of fuel ratio on the performance of combustion synthesized bifunctional cobalt oxide catalysts for fuel cell application. Int J Hydrogen Energy 2019;44:436–45. https://doi.org/10.1016/j.ijhydene.2018.02.111.

[35] Hao Y, Jin J, Jin H. Thermodynamic analysis of isothermal CO2 splitting and CO2-H2O co-splitting for solar fuel production. Appl Therm Eng 2020;166:113600. https://doi.org/10.1016/j.applthermaleng.2019.04.010.

[36] Ashok A, Kumar A, Bhosale RR, Saleh MAH, Ghosh UK, Al-Marri M, et al. Cobalt oxide nanopowder synthesis using cellulose assisted combustion technique. Ceram Int 2016;42:12771–7. https://doi.org/10.1016/j.ceramint.2016.05.035.

[37] Bhosale RR, Kumar A, AlMomani F, Ghosh U, Dardor D, Bouabidi Z, et al. Solar co-production of samarium and syngas via methanothermal reduction of samarium sesquioxide. Energy Convers Manag 2016;112:413–22. https://doi.org/10.1016/j.enconman.2016.01.032.

[38] Steinfeld A. Solar hydrogen production via a two-step water-splitting thermochemical cycle based on Zn/ZnO redox reactions. Int J Hydrogen Energy 2002;27:611–9. https://doi.org/10.1016/S0360-3199(01)00177-X.

[39] Bhosale RR, Alxneit I, van den Broeke LLP, Kumar A, Jilani M, Gharbia SS, et al. Sol-Gel Synthesis of Nanocrystalline Ni-Ferrite and Co-Ferrite Redox Materials for Thermochemical Production of Solar Fuels. MRS Proceedings 2014;1675:203–8. https://doi.org/10.1557/opl.2014.866.

[40] Diver RB, Miller JE, Allendorf MD, Siegel NP, Hogan RE. Solar Thermochemical Water-Splitting Ferrite-Cycle Heat Engines. J Sol Energy Eng 2008;130. https://doi.org/10.1115/1.2969781.

[41] Bhosale RR, Kumar A, Ashok A, Sutar P, Takalkar G, Khraisheh M, et al. La-Based Perovskites as Oxygen-Exchange Redox Materials for Solar Syngas Production. MRS Adv 2017;2:3365–70. https://doi.org/10.1557/adv.2017.300.
[42] Bare ZJL, Morelock RJ, Musgrave CB. A Computational Framework to Accelerate the Discovery of Perovskites for Solar Thermochemical Hydrogen Production: Identification of Gd Perovskite Oxide Redox Mediators. Adv Funct Mater 2022;32:2200201. https://doi.org/10.1002/adfm.202200201.
[43] Takalkar G, Bhosale RR. Solar thermocatalytic conversion of CO2 using PrxSr(1−x)MnO3−δ perovskites. Fuel 2019;254:115624. https://doi.org/10.1016/j.fuel.2019.115624.
[44] Li S, Wheeler VM, Kumar A, Venkataraman MB, Muhich CL, Hao Y, et al. Thermodynamic Guiding Principles for Designing Nonstoichiometric Redox Materials for Solar Thermochemical Fuel Production: Ceria, Perovskites, and Beyond. Energy Technology 2022;10:2000925. https://doi.org/10.1002/ente.202000925.
[45] Bhosale RR. Thermodynamic analysis of Ni-ferrite based solar thermochemical H2O splitting cycle for H2 production. Int J Hydrogen Energy 2019;44:61–71. https://doi.org/10.1016/j.ijhydene.2018.03.145.
[46] Scheffe JR, Li J, Weimer AW. A spinel ferrite/hercynite water-splitting redox cycle. Int J Hydrogen Energy 2010;35:3333–40. https://doi.org/10.1016/j.ijhydene.2010.01.140.
[47] Takalkar G, Bhosale R, AlMomani F. Combustion synthesized A0.5Sr0.5MnO3-δ perovskites (where, A = La, Nd, Sm, Gd, Tb, Pr, Dy, and Y) as redox materials for thermochemical splitting of CO2. Appl Surf Sci 2019;489:80–91. https://doi.org/10.1016/j.apsusc.2019.05.284.
[48] Jacot R, Moré R, Michalsky R, Steinfeld A, Patzke GR. Trends in the phase stability and thermochemical oxygen exchange of ceria doped with potentially tetravalent metals. J Mater Chem A 2017;5:19901–13. https://doi.org/10.1039/C7TA04063F.
[49] Luciani G, Landi G, Di Benedetto A. Syngas Production Through H2O/CO2 Thermochemical Splitting Over Doped Ceria-Zirconia Materials. Front Energy Res 2020;8. https://doi.org/10.3389/fenrg.2020.00204.
[50] Bhosale RR, Shende RV, Puszynski JA. Sol-Gel Synthesis of Ferrite Foam Materials for H2 Generation from Water-Splitting Reaction. Proceedings of NSTI–Nanotech Conference & Expo., 2010, p. 368–71.
[51] Tamaura Y, Hasegawa N, Kojima M, Ueda Y, Amano H, Tsuji M. Water splitting with the Mn(II)-ferrite–CaO–H2O system at 1273K. Energy 1998;23:879–86. https://doi.org/10.1016/S0360-5442(98)00030-9.
[52] Dardor D, Bhosale RR, Gharbia S, Kumar A, AlMomani FA. Solar carbon production via thermochemical ZnO/Zn carbon dioxide splitting cycle. Journal of Emerging Trends in Engineering and Applied Sciences 2016;7:129–35.
[53] Chambon M, Abanades S, Flamant G. Solar thermal reduction of ZnO and SnO2: Characterization of the recombination reaction with O2. Chem Eng Sci 2010;65:3671–80. https://doi.org/10.1016/j.ces.2010.03.005.

[54] Bhosale RR, Shende R V, Puszynski JA. H 2 Generation From Thermochemical Water-Splitting Using Sol-Gel Synthesized Zn/Sn/Mn-doped Ni-Ferrite. International Review of Chemical Engineering (IRECHE) 2010;2.

[55] Kaneko H, Gokon N, Hasegawa N, Tamaura Y. Solar thermochemical process for hydrogen production using ferrites. Energy 2005;30:2171–8. https://doi.org/10.1016/j.energy.2004.08.020.

[56] Shende RV, Puszynski JA, Opoku MK, Bhosale RR. Synthesis of novel ferrite foam material for water-splitting application. Proceedings of NSTI Nanotech conference & expo., 2009, p. 201–4.

[57] Scheffe JR, McDaniel AH, Allendorf MD, Weimer AW. Kinetics and mechanism of solar-thermochemical H2 production by oxidation of a cobalt ferrite–zirconia composite. Energy Environ Sci 2013;6:963. https://doi.org/10.1039/c3ee23568h.

[58] Bhosale RR, Shende R V, Puszynski JA. H 2 Generation from Thermochemical Water-Splitting Using Sol-Gel Derived Ni-Ferrite. Journal of Energy and Power Engineering 2010;4.

[59] Mostrou S, Büchel R, Pratsinis SE, van Bokhoven JA. Improving the ceria-mediated water and carbon dioxide splitting through the addition of chromium. Appl Catal A Gen 2017;537:40–9. https://doi.org/10.1016/j.apcata.2017.03.001.

[60] Bhosale R, Khadka R, Puszynski J, Shende R. H2 generation from two-step thermochemical water-splitting reaction using sol-gel derived SnxFeyOz. Journal of Renewable and Sustainable Energy 2011;3. https://doi.org/10.1063/1.3659684.

[61] Agrafiotis C, Zygogianni A, Pagkoura C, Kostoglou M, Konstandopoulos AG. Hydrogen production via solar-aided water splitting thermochemical cycles with nickel ferrite: Experiments and modeling. AIChE Journal 2013;59:1213–25. https://doi.org/10.1002/aic.13882.

[62] Bhosale RR, Kumar A, AlMomani F, Ghosh U, Sutar P, Takalkar G, et al. Effectiveness of Ni incorporation in iron oxide crystal structure towards thermochemical CO2 splitting reaction. Ceram Int 2017;43:5150–5. https://doi.org/10.1016/j.ceramint.2017.01.034.

[63] Haeussler A, Julbe A, Abanades S. Investigation of reactive perovskite materials for solar fuel production via two-step redox cycles: Thermochemical activity, thermodynamic properties and reduction kinetics. Mater Chem Phys 2022;276:125358. https://doi.org/10.1016/j.matchemphys.2021.125358.

[64] Bhosale RR, Sutar P, Kumar A, AlMomani F, Ali MH, Ghosh U, et al. Solar hydrogen production via erbium oxide based thermochemical water splitting cycle. Journal of Renewable and Sustainable Energy 2016;8:034702. https://doi.org/10.1063/1.4953166.

[65] Bhosale RR. Thermodynamic efficiency analysis of zinc oxide based solar driven thermochemical H2O splitting cycle: Effect of partial pressure of O2, thermal reduction and H2O splitting temperatures. Int J Hydrogen Energy 2018;43:14915–24. https://doi.org/10.1016/j.ijhydene.2018.06.074.

[66] Jacot R, Naik JM, Moré R, Michalsky R, Steinfeld A, Patzke GR. Reactive stability of promising scalable doped ceria materials for thermochemical two-step CO_2 dissociation. J Mater Chem A Mater 2018;6:5807–16. https://doi.org/10.1039/C7TA10966K.

[67] Bhosale RR, Takalkar GD. Nanostructured co-precipitated Ce0.9Ln0.1O2 (Ln = La, Pr, Sm, Nd, Gd, Tb, Dy, or Er) for thermochemical conversion of CO2. Ceram Int 2018;44:16688–97. https://doi.org/10.1016/j.ceramint.2018.06.096.
[68] Bhosale RR, Kumar A, AlMomani F, Alxneit I. Propylene oxide assisted sol–gel synthesis of zinc ferrite nanoparticles for solar fuel production. Ceram Int 2016;42:2431–8. https://doi.org/10.1016/j.ceramint.2015.10.043.
[69] Dey S, Rao CNR. Splitting of CO $_2$ by Manganite Perovskites to Generate CO by Solar Isothermal Redox Cycling. ACS Energy Lett 2016;1:237–43. https://doi.org/10.1021/acsenergylett.6b00122.
[70] Bhosale RR, Kumar A, AlMomani F, Ghosh U, Khraisheh M. A comparative thermodynamic analysis of samarium and erbium oxide based solar thermochemical water splitting cycles. Int J Hydrogen Energy 2017;42:23416–26. https://doi.org/10.1016/j.ijhydene.2017.03.172.
[71] Padella F, Alvani C, Barbera A La, Ennas G, Liberatore R, Varsano F. Mechanosynthesis and process characterization of nanostructured manganese ferrite. Mater Chem Phys 2005;90:172–7. https://doi.org/10.1016/j.matchemphys.2004.10.033.
[72] Bhosale R, Kumar A, AlMomani F. Solar Thermochemical Hydrogen Production via Terbium Oxide Based Redox Reactions. International Journal of Photoenergy 2016;2016:1–9. https://doi.org/10.1155/2016/9727895.
[73] Chueh WC, Falter C, Abbott M, Scipio D, Furler P, Haile SM, et al. High-Flux Solar-Driven Thermochemical Dissociation of CO $_2$ and H $_2$ O Using Nonstoichiometric Ceria. Science (1979) 2010;330:1797–801. https://doi.org/10.1126/science.1197834.
[74] Takalkar GD, Bhosale RR, Kumar A, AlMomani F, Khraisheh M, Shakoor RA, et al. Transition metal doped ceria for solar thermochemical fuel production. Solar Energy 2018;172:204–11. https://doi.org/10.1016/j.solener.2018.03.022.
[75] Roeb M, Säck J-P, Rietbrock P, Prahl C, Schreiber H, Neises M, et al. Test operation of a 100kW pilot plant for solar hydrogen production from water on a solar tower. Solar Energy 2011;85:634–44. https://doi.org/10.1016/j.solener.2010.04.014.
[76] Bhosale RR, Shende R V., Puszynski JA. Sol-Gel Derived NiFe $_2$ O $_4$ Modified with ZrO $_2$ for Hydrogen Generation from Solar Thermochemical Water-Splitting Reaction. MRS Proceedings 2012;1387:mrsf11-1387-e09-07. https://doi.org/10.1557/opl.2012.772.
[77] Fresno F, Fernández-Saavedra R, Belén Gómez-Mancebo M, Vidal A, Sánchez M, Isabel Rucandio M, et al. Solar hydrogen production by two-step thermochemical cycles: Evaluation of the activity of commercial ferrites. Int J Hydrogen Energy 2009;34:2918–24. https://doi.org/10.1016/j.ijhydene.2009.02.020.

87a Steam Reforming of Model Biomass Gasification Tar over Ni-Based Carbon Supported Catalyst

Sang Jun Yoon, Sung Jin Park, Tae-Young Mun, Jihong Moon

Tar in biomass gasification synthetic gas is one of the factors that have the greatest influence on the stable operation of biomass gasification system and the utilization of synthetic gas. In this study, the catalytic steam reforming was evaluated to convert tar into hydrogen. Toluene, which accounts for a large content of biomass tar, was used as a model biomass tar. One of the biggest problems in catalytic steam reforming of hydrocarbon materials is the deactivation of catalyst due to carbon deposition. To overcome this, carbon support with high coking resistance was applied. A catalyst was prepared using activated carbon, carbon black (CB), coal, and nanocellulose as a support, Ni as an active metal, and Mg, K, and Ce as a promoter. Under constant conditions with a toluene concentration of 30 g/Nm^3, the catalyst type, reaction temperature, steam/carbon molar ratio, and gas hourly space velocity effects were studied. In general, CB support improved catalyst reactivity because of its high thermal stability, mesoporous structure, and specific surface area, and the addition of a promoter to the Ni/CB catalyst increases the reactivity at low temperatures. All catalysts used in this study achieved H_2 yield and carbon conversion of more than 70% and 87% without a reduction of catalyst before the reaction. The NiK/CB catalyst was most effective with a 100% carbon conversion rate and hydrogen yield at 550°C. The NiMg/CB catalyst confirmed stable reactivity without deactivation by coking for 50 hours.

Acknowledgment

This work was supported by the ERC Center funded by the National Research Foundation of Korea (NRF-2022R1A5A1033719).

87b Effects on Ammonia Injection Directions for Coal-Ammonia Co-Firing in a Circulating Fluidized Bed Combustion Test Rig

Seong-Ju Kim, Sung Jin Park, Sang Jun Yoon, Tae-Young Mun

Ammonia (NH_3) co-firing both pulverized coal boiler and circulating fluidized bed boiler is a promising technology for reducing greenhouse gas emissions, important to achieve 2030 NDC(Nationally determined contribution) and 2050 carbon neutrality. By this situation, researches on NH_3 co-firing are conducting since 2021 in South Korea. In order to be close to commercialization of NH_3 co-firing, optimization of ammonia injection method and the comprehensive combustion characteristics, including pollutant emission, combustion efficiency, and ash properties on coal-ammonia co-firing should be investigated. Therefore, in this work, NH_3 co-firing characteristics on ammonia injection directions in dense bed zone were observed to optimize NH_3 supply way during coal–NH_3 co-firing ratio (21.2% as thermal input basis) in a pilot-scale circulating fluidized bed (CFB) combustion test rig. As a result, compared with only coal-fired combustion, all NH_3 co-firing operations on variation of ammonia injection directions simultaneously reduced NO and CO_2 emission, achieving the highest combustion efficiency without ash-related problems.

87c Techno Feasibility Analysis for Steam-to-Hot Water Conversion for West Virginia University, Morgantown Campus.

Sai Kiran Yerravally, Daniel Lemasters, Thomas J. Rosson, Srinivas Palanki, Nagasree Garapati

The Low-temperature Geothermal Play Fairway Analysis of the Appalachian Basin (GPFA-AB) estimated a region of elevated heat flows in northcentral West Virginia (Cornell University 2017). The Morgantown campus of West Virginia University (WVU) is located within this region and offers a desirable and unique combination of critical factors necessary to develop a deep direct-use geothermal system in the eastern United States. Based on our initial feasibility analysis study, it is found that the geothermal DDU system at WVU is feasible to replace the existing system. In this work, a techno-economic analysis to evaluate the conversion of the existing steam-based system to a geothermal hot water-based system will be performed.

Currently, individual building energy loads are estimated by collecting weekly steam flow rates using existing meters and new ultra-sonic clamp-on meters. Assessment of existing building equipment providing heating and cooling across the campus is performed. The suitability of current building-level steam equipment for hot water retrofit will be established, and costs for retrofit or replacement will be calculated.

Acknowledgment: This material is based upon work supported by the U.S. DOE's Office of EERE under the GTO, under Award Number DE- EE0009597.

87e Stabilizing Sodium Ion Transfer at the Nasicon Solid-State Electrolyte and Metallic Na Anode Interface By Nanoscale Metal Oxide Coating

Baiheng Li, Ian Baker, Weiyang Li

Sodium (Na) ion batteries (SIBs) are widely regarded as a promising solution to large-scale electrical energy storage demands due to the natural abundance of metal and its high energy density of 1166 mAh/g as a metallic anode [1]. In order to advance the practicality of SIBs, we must harness the highly reactive nature of Na metal anode, and by employing solid-state electrolytes we can eliminate the safety risk caused by the possible reactions between Na metal and conventional liquid electrolytes [2].

Na^+ superionic conductor ($Na_3Zr_2Si_2PO_{12}$, NASICON) is a ceramic-type solid-state electrolyte discovered by Goodenough et. al. in 1970s [3]. Its relatively high ionic conductivity at room temperature and exceptional thermal and chemical stability allows it to become one of the most studied solid-state electrolytes for Na batteries. However, the inherent poor wettability between NASICON and Na metal anode leads to high interfacial impedance and localized Na^+ flux, which are detrimental to the long-term cyclability of solid-state batteries. Therefore, various approaches have been attempted to reduce interfacial resistance and promote uniform plating and stripping of Na+, with one example being the introduction of an artificial interlayer [4].

In this work, we aim to investigate nanoscale metal oxide coatings' effect on stabilizing Na^+ transfer in a systematic manner. Metal oxide coatings have great compatibility with NASICON, and once proven effective, their deposition can be a scalable and cost-effective method in the production of solid-state Na metal batteries. Atomic layer deposition (ALD) is employed to deposit various metal oxides of tunable thicknesses onto the NASICON surface, and after optimizing the composition and thickness of the coating layer, symmetric cells can demonstrate critical current densities (CCD) up to 12 mA/cm^2, which is 15 times higher than the CCD of 0.8 mA/cm^2 seen with the unmodified NASICON. In addition, the metal oxide coated NASICON allows stable plating/stripping cycling performance at a current density of 1.6 mA/cm^2 for over 300 cycles and 600 hours. In order to reveal the mechanism behind the performance improvement, the crystallinities and chemical compositions of uncoated and coated NASICON samples were studied using X-ray diffraction (XRD) and X-ray photoelectron spectroscopy (XPS) both as fabricated and after an extended period of cycling. The grain size and surface morphology are studied with scanning electron microscopy (SEM), and cross-sectional samples for transmission electron microscopy (TEM) are obtained using focused-ion beam (FIB) in order to investigate the microstructures formed at the interface of the metal oxide coating and NASICON. In-situ electrochemical cells compatible with X-ray micro-computed tomography (micro-CT) are fabricated for direct observation of the morphological evolution along with localized Na+ transfer in uncoated NASICON and uniform Na+ transfer in NASICON with nanoscale metal oxide coating.

References

[1] Tang, B.; Jaschin, P. W.; Li, X.; Bo, S.-H.; Zhou, Z. Critical interface between inorganic solid-state electrolyte and sodium metal. Mater. Today 2020, 41, 200– 218, DOI: 10.1016/j.mattod.2020.08.016

[2] Manthiram, A.; Yu, X.; Wang, S. Lithium Battery Chemistries Enabled by Solid-State

Electrolytes. Nat. Rev. Mater. 2017, 2, 16103, DOI: 10.1038/natrevmats.2016.103

[3] Goodenough, J. B.; Hong, H.Y.-P.; Kafalas, J. A. Fast Na+ -ion transport in skeleton structures. Mater. Res. Bull. 1976, 11, 203, DOI:10.1016/0025-5408(76)90077-5.

[4] Matios, E.; Wang, H.; Wang, C. L.; Hu, X. F.; Lu, X.; Luo, J. M.; Li, W. Y. Graphene Regulated Ceramic Electrolyte for Solid-State Sodium Metal Battery with Superior Electrochemical Stability. ACS Appl. Mater. Interfaces 2019, 11, 5064– 5072, DOI: 10.1021/acsami.8b19519

87f Biogas Production from Co-Digesting House Hold Vegetable Waste with the Industrial Waste Water Using Expanded Granular Sludge Bed Anaerobic Digestion (AD) System

Jackson Oglesby, Sarah Riley, Frank Reinsch, Jaden Johnson, Vena Prudent, Justin Penn, Haider Al-Rubaye, Joseph Smith

AD reactors are an established technology that are used to treat organic waste (liquid or solid carbon content wastes). The product of this process is biogas, which consists of methane (50-75%) and carbon dioxide (25-50%). The generated gas could be used for generating electricity, heating purposes, or upgrading to the same quality of natural gas. In this study, the team wanted to evaluate the efficacy of using vegetable waste from on-campus dining services as the primary feedstock for the AD process. Solid vegetable waste was gathered and blended. A benchtop scale of the process was performed.

87g Insight into Electrochemical Batch Reactor for Phosphorus Recovery Using Mathematical Modeling

Ardavan Zanganeh, Kody Wolfe, Damilola Daramola

Recovery of phosphorus from wastewater is a crucial step in managing phosphorus resources and reducing environmental pollution. Among various conventional and emerging approaches, electrochemical reactors have gained considerable attention as a cost-effective and efficient method for recovering phosphorus from wastewater (Bagastyo et al., 2022), by conversion to a solid form, such as struvite, which can be re-used as a fertilizer.

Introduction of mathematical continuum modeling into the study of electrochemical phosphorus recovery phenomena could play a crucial role in the design and optimization of reactors for this emerging recovery approach. It allows researchers to predict the behavior of the system under different conditions and to identify critical points in the reactor that can affect its performance. Mathematical models can also help reduce the cost and time required for experimentation and enable the analysis of phenomena that are difficult to observe experimentally.

In this study, a 2-D mathematical model of an electrochemical reactor was developed using COMSOL Multiphysics software. The model was used to investigate the hydrodynamics of the reactor, including impellers, electrochemical reactions, mass transfer of electroactive species, and phosphorus recovery. The model was solved numerically using the finite element method. The hydrodynamics of the reactor were investigated using the k-epsilon turbulence model, which is widely employed in literature to simulate the fluid flow and mixing behavior of the reactants. The electrochemical reactions that occur in the reactor were modeled using the Butler-Volmer equation at the cathode and the Tafel equation at the anode. The Butler-Volmer equation describes kinetics, while accounting for the concentration of electroactive species, the exchange current density, and the overpotential. On the other hand, the Tafel equation describes kinetics when the reaction rate is not controlled by the transport of reactants to the electrode surface. Boundary conditions for the potentials were a grounded counter electrode (anode) and applying -1.3 vs Ag/AgCl(sat.) to the working electrode (cathode). Mass transfer of electroactive species, such as phosphorus and oxygen, was modeled using Fick's second law of diffusion and the convective transport equation. The model also considered the migration of ions due to the electric field, which can affect the distribution of species inside the reactor.

Preliminary results indicate that although the impellers can create turbulence and affect the local conditions around the electrodes, the backside of the electrodes is weakly affected by impeller rotation (Figure 1) which may adversely affect the mass transfer rate and electrochemical reactions. These results will be integrated with chemical and electrochemical reactions and then presented at the meeting, thus providing insight into the reactor behavior and options for optimization.

Figure 1: Flow field in a batch electrochemical reactor with impeller speed of 200 RPM, and at t = 21 s

Reference:

Bagastyo, A. Y., Anggrainy, A. D., Khoiruddin, K., Ursada, R., Warmadewanthi, I., & Wenten, I. G. (2022). Electrochemically-driven struvite recovery: Prospect and challenges for the application of magnesium sacrificial anode. *Separation and Purification Technology*, *288*, 120653. https://doi.org/10.1016/j.seppur.2022.120653

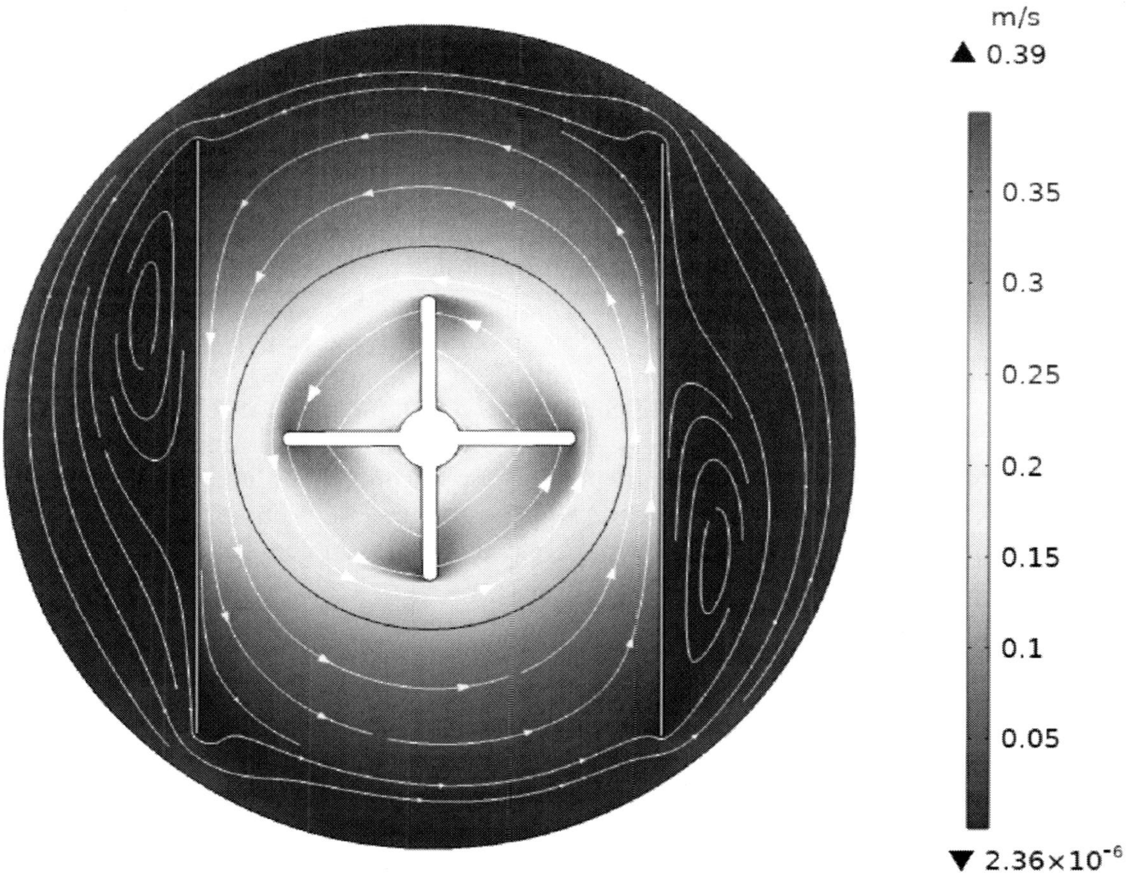
m/s
▲ 0.39
0.35
0.3
0.25
0.2
0.15
0.1
0.05
▼ 2.36×10⁻⁶

87h Optimization of Capacitance in Supercapacitors By Constructing an Experimentally Validated Hybrid Artificial Neural Networks-Genetic Algorithm Framework

Betul Uralcan

Supercapacitors are high power electrochemical energy storage systems that are attractive candidates for use in high power applications. Yet, their widespread adoption has been restricted due to their relatively low energy density. Improving the energy storage performance of supercapacitors is linked to rationally optimizing the key descriptors that affect capacitance. This work presents a systematic approach based on a hybrid artificial neural network (ANN) and genetic algorithm (GA) integrated with the Big M method to efficiently and rationally design carbon-based supercapacitors with improved energy storage performance. By performing structural and electrochemical characterization on systems we fabricate, we experimentally validate the robustness and generalizability of the developed ANN-GA framework. This study takes a step towards the rational design of supercapacitors by implementing the hybrid ANN-GA framework as an optimization tool to provide guidelines for rationally tuning material properties and operational conditions for improved capacitance.

87i In silico Demonstration of Fast Anhydrous Proton Conduction on Graphanol

Siddarth Achar, Leonardo Bernasconi, Karl Johnson

Development of new materials capable of conducting protons in the absence of water is crucial for improving the performance, reducing the cost, and extending the operating conditions for proton exchange membrane fuel cells. We present detailed atomistic simulations showing that graphanol (hydroxylated graphane) will conduct protons anhydrously with very low diffusion barriers. We developed a deep learning potential (DP) for graphanol that has near-density functional theory accuracy but requires a very small fraction of the computational cost. We used our DP to calculate proton self-diffusion coefficients as a function of temperature, to estimate the overall barrier to proton diffusion, and to characterize the impact of thermal fluctuations as a function of system size. We propose and test a detailed mechanism for proton conduction on the surface of graphanol. We show that protons can rapidly hop along Grotthuss chains containing several hydroxyl groups aligned such that hydrogen bonds allow for conduction of protons forward and backward along the chain without hydroxyl group rotation. Long-range proton transport only takes place as new Grotthuss chains are formed by rotation of one or more hydroxyl groups in the chain. Thus, the overall diffusion barrier consists of a convolution of the intrinsic proton hopping barrier and the intrinsic hydroxyl rotation barrier. Our results provide a set of design rules for developing new anhydrous proton conducting membranes with diffusion barriers lower than existing systems.

87j Ageing Effects of Catalyst Ink for Polymer Electrolyte Membrane Fuel Cells

Mario Kircher, Wai Yee Koo, Michaela Roschger, Viktor Hacker

The major contribution, about 90 %, of the energy demand in the transport sector is up to date provided by petroleum products [1]. A viable way to reduce greenhouse gas emissions in this sector is to use hydrogen as a fuel for polymer electrolyte membrane fuel cell (PEMFC) vehicles. The widespread commercialisation of fuel cell electric vehicles demands an increase in reliability and durability as well as a reduction in costs [2,3]. The performance-determining core unit of a PEMFC, the membrane electrode assembly, consists of a polymer electrolyte membrane which is covered with anodic and cathodic catalyst layers (CLs) on both sites, respectively. CLs are formed by applying catalyst inks (CI) onto the PEM, with different application methods, such as spray coating or inkjet printing. Most commonly, the CI consists of a catalyst (e.g., platinum), an ionomer as binder, as well as an alcohol-water mixture as solvent. The electrochemical reactions take place in the CI and consequently the CI is vital for the performance of a PEMFC. Unfortunately, ageing effects during CI storage can cause activity losses and performance degradation.

In this work, we demonstrate CI ageing degradation effects under various aspects over a long storage time of up to three weeks. The composition of the CI (Pt/C, Nafion, isopropanol, water) is kept constant for all measurements and based on in-house experience. Sedimentation studies reveal a significant change in concentration of the components over time despite the overall adequate stability of dispersion. Degradation products formed by the oxidation of isopropanol are qualified and quantified. Results show their non-negligible presence even in freshly prepared CI. Half and fuel cell measurements are used to determine the effect of storage on the catalytic activity and hence, the cell performance. Results could be used for considerations about CI preparation and storage in small and large-scale PEMFC production e.g., for the reduction of waste or increased PEMFC performance.

References

1. S. Energy Information Administration. *Use of energy explained: Energy use for transportation.* 2021. URL: https://www.eia.gov/energyexplained/use-of-energy/transportation.php (accessed: 11/04/2022)

2. Iain Staffell et al. "The role of hydrogen and fuel cells in the global energy system". In: *Energy & Environmental Science* 12 (2019), pp. 463 – 491. DOI: 1039/C8EE01157E

3. Gregory Trencher. "Strategies to accelerate the production and diffusion of fuel cell electric vehicles: Experiences from California". In: *Energy Reports* 6 (2020), pp. 2503 - 2519. DOI: 1016/j.egyr.2020.09.008

87k Electrode Development in Alkaline Direct Ethanol Fuel Cells

Michaela Roschger, Sigrid Wolf, Boštjan Genorio, Selestina Gorgieva, Viktor Hacker

Climate change and electricity consumption of the population have led to increasing interest in technologies that generate electricity from renewable sources. The use of sustainable ethanol in alkaline direct ethanol fuel cells (ADEFC) offers the advantages of easy storage and transport of the liquid fuel, low toxicity, environmental friendliness and robustness of these cells. Moreover, the alkaline environment in the ADEFC promotes the electrochemical reactions at the cathode and anode, and allows the use of platinum-free catalysts and low-cost anion exchange membranes. Besides challenges such as the incomplete conversion of ethanol (C-C cleavage issues), mixed potentials due to ethanol crossover and the durability of membranes, a high catalyst layer utilization of the electrodes needs to be achieved before commercialization.[1-4]

In this study, the effect of metal/carbon ratio of the anode catalyst and the influence of the layer thickness of both electrodes, anode and cathode, were investigated. Hence, thin-film rotating disk electrode (RDE) and half-cell gas diffusion electrode (GDE) experiments, as well as single cell tests were performed. Therefore, catalyst with different active material contents, electrodes with different layer thicknesses, and membrane electrode assemblies consisting of different catalysts and loadings were fabricated. The half-cell GDE and single-cell tests were conducted at different operating temperatures and conditions to determine the influence. Moreover, the produced electrodes were analysed with scanning electron microscopy. The reduction of metal content of the anodic catalysts resulted in more accessible metal particles and thus in a higher activity and performance and moreover in a cost reduction.[1] Furthermore, it could be shown that with half-cell GDE measurements, realistic reaction conditions, such as higher currents for the oxygen reduction reaction in comparison to RDE experiments, can be realized. The layer thickness influences the crossover through the membrane, as well as the catalyst layer utilization. High maximum power densities of 120 mW cm^{-2} were obtained by determining the optimum layer thicknesses for the cathodic and anodic electrodes with Pt-free catalysts.[4] In addition, the investigations show potential further improvements of the electrodes for the commercialisation of the ADEFC.

The financial support from the Austrian Science Fund (FWF I 3871-N37) is acknowledged.

References

1. Roschger, S. Wolf, B. Genorio, V. Hacker. Effect of PdNiBi Metal Content: Cost Reduction in Alkaline Direct Ethanol Fuel Cells, Sustainability 14 (2022) 15485 (https://doi.org/https://doi.org/10.3390/su142215485)

2. Wolf, M. Roschger, B. Genorio, M. Kolar, D. Garstenauer, B. Bitschnau, V. Hacker. Ag-MnxOy on Graphene Oxide Derivatives as Oxygen Reduction Reaction Catalyst in Alkaline Direct Ethanol Fuel Cells, Catalysts 12 (2022) 780 (https://doi.org/10.3390/catal12070780)

3. Gorgieva, A. Osmić, S. Hribernik, M. Božič, J. Svete, V. Hacker, S. Wolf, B. Genorio. Efficient chitosan/nitrogen-doped reduced graphene oxide composite membranes for direct alkaline ethanol fuel cells, International Journal of Molecular Sciences 22 (2021) 1740 (https://doi.org/10.3390/ijms22041740)

4. Roschger, S. Wolf, K. Mayer, A. Billiani, B. Genorio, S. Gorgieva, V. Hacker. Influence of the electrocatalyst layer thickness on alkaline DEFC performance, Sustainable Energy and Fuels 7 (2023)

1093–1106 (https://doi.org/10.1039/d2se01729f)

87l Multi-Physics Simulations and Bayesian Optimisation of Flow in Alkaline Water Electrolyzers

Morgan Kerhouant, Thomas Abadie, Raj Venuturumilli, Andre Nicolle, Omar Matar

Hydrogen will play a critical role in the energy transition, with the 2021 UK hydrogen strategy aiming to replace 20% of the UK's natural gas usage with low carbon hydrogen by 2030 [1]. With the potential to replace fossil fuels and decarbonise sectors such as energy, transport, and heating, reducing the cost of green hydrogen is key to achieving net zero by 2050. Alkaline water electrolysers produce hydrogen and oxygen through electrolysis of water and are commonly employed for industrial-scale production of low carbon hydrogen. Operating at high current densities leads to enhanced hydrogen production but reduced cell efficiency, partly due to ohmic losses from the gas-liquid flow. Bubble coverage of the electrode surface and reduced effective electrolyte conductivity from increased void fraction are major contributors to these losses.

To gain further insights into the flow in electrochemical cells and optimise the design and operation of AWEs, we perform three-dimensional transient multi-physics simulations with the OpenFOAM libraries. We simulate the bubbly flow with a multi-fluid Eulerian model and consider bubble coalescence and polydispersity through population balance modelling (PBM). The electrical current distribution in the electrolyte governs the gas production rate at each electrode through Faraday's law. We solve the electrical potential and employ an iterative method to impose boundary conditions that satisfy the non-linear relationship between current density and Butler-Volmer kinetics at the electrodes [2]. We model heat transfer arising from Joule heating in the electrolyte and activation overpotentials at the electrodes. We compare our results with experimental work, and study free convection [3] and forced convection configurations with diaphragms [4] over a range of current densities and operating conditions. We examine the assumptions commonly employed in multiphase modelling of AWEs, including constant bubble diameter, no coalescence, isothermal conditions and uniform current density distribution, and analyse the contributions to flow and cell performance predictions from flow, electrochemistry, and heat transfer models.

Simulations results suggest a high sensitivity to various modelling parameters, particularly within PBM and momentum-coupling terms. Several of these parameters, including bubble size, distribution and turbulent dispersion coefficients are obtained through experimental methods. However, significant uncertainty is present in some of these measurements, with large variations in published bubble diameters, and unknown coalescence rates and turbulent dispersion coefficients for bubbles in electrolyte solutions, potentially resulting in significant errors in simulation results and inaccurate predictions. To avoid large scale parametric studies which are computationally expensive for 3D high resolution simulations, we employ a Bayesian optimisation and surrogate modelling approach, allowing us to search the parameter space and optimise our model in a significantly lower number of iterations and at lower computational costs. Our OpenFOAM solver is coupled to the scikit-optimise library and runs autonomously to determine optimal input parameters by minimising an objective function based on error between predicted and experimental gas distributions.

We obtain good agreement with experimental work over the range of operating current densities (500-6250 Am^{-2}). Bubble diameter has a large impact on the flow, with larger diameter and low turbulent dispersion coefficients resulting in low volume fractions near the electrode, while small bubbles lead to insufficient gas dispersion, as displayed in 1a. PBM coefficients derived from the Bayesian optimisation approach in 1b, with reduced coalescence efficiency to account for the electrolyte solutions, lead to improved gas distribution. Furthermore, we find that under the parameters investigated for the experimental lab-scale set-up, Joule heating plays an insignificant role on the flow, and results in temperature increases of up to 1.1◦C at high current densities, as illustrated in figure 1c.

Acknowledgements: This work is supported by EPSRC Programme Grant PREMIERE (EP/T000414/1). The authors would like to acknowledge funding for M.K through an ICASE studentship co-funded by

EPSRC and bp, as well as technical support from bp through the bp-ICAM.

References

1. [1] Department for BEIS, "UK Hydrogen Strategy," Tech. Rep., 2021.

2. [2] A. N. Colli and H. H. Girault, "Compact and General Strategy for Solving Current and Potential Distribution in Electrochemical Cells Composed of Massive Monopolar and Bipolar Electrodes," Journal of The Electrochemical Society, vol. 164, no. 11, pp. E3465–E3472, 6 2017.

3. [3] P. Boissonneau and P. Byrne, "Experimental investigation of bubble-induced free con- vection in a small electrochemical cell," Journal of Applied Electrochemistry, vol. 30, no. 7, pp. 767–775, 2000.

4. [4] H. Riegel, J. Mitrovic, and K. Stephan, "Role of mass transfer on hydrogen evolution in aqueous media," Journal of Applied Electrochemistry, vol. 28, no. 1, pp. 10–17, 1 1998.

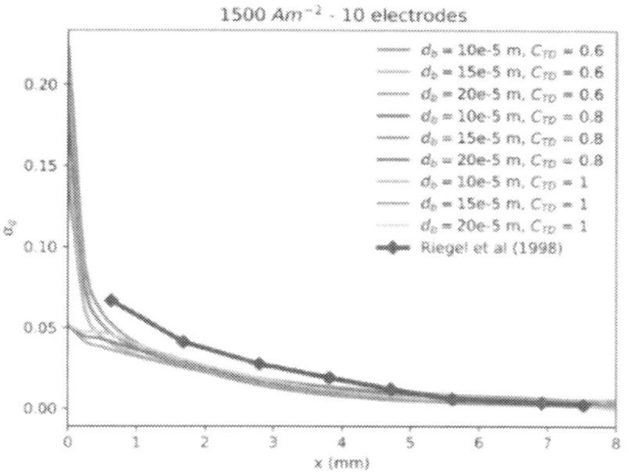

(a) Gas volume fraction across cathode channel at y=0.5 m.

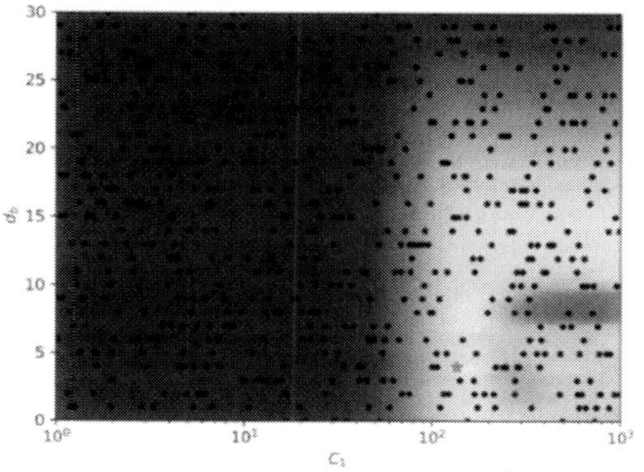

(b) Contour plot of PBM parameter objective function. Sampling points are displayed in black and optimal point is in red.

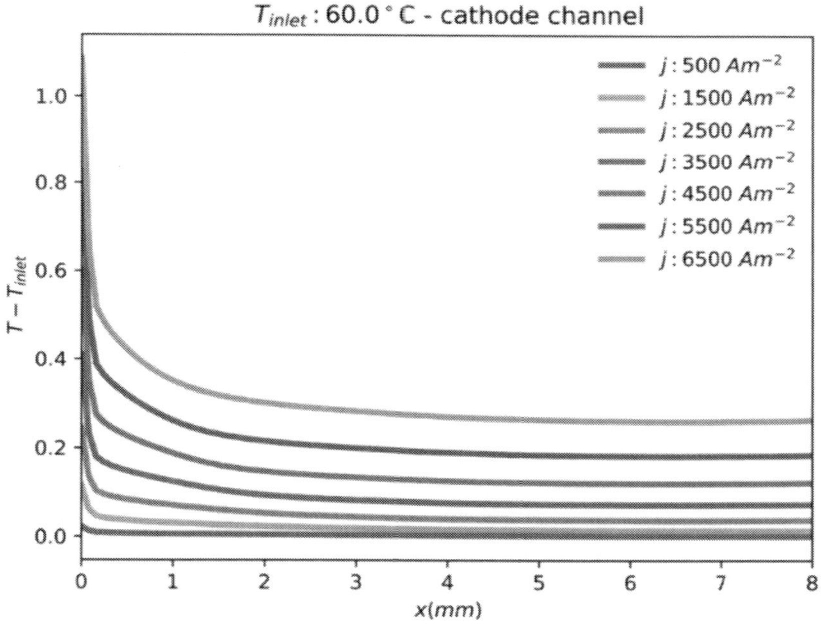

(c) Temperature increase in cathode channel.

Figure 1: Numerical modelling and Bayesian optimisation results.

87m Ion and Water Transport in Polynorbornene-Based Thin Film Membranes

Ge Sun, Zhongyang Wang, Abhishek Sharma, Shrayesh Patel, Paul F. Nealey, Juan J. de Pablo

Insight into ion transport within Anion Exchange Membranes (AEMs) is critical for crafting efficient polymer electrolytes, integral to fuel cells and numerous electrochemical technologies. One significant hurdle in characterizing ion transport in AEMs is the lack of an inclusive methodology that bridges macroscopic experimental data with simulation results and disentangles ion, water, and polymer dynamics at varying hydration states on a molecular level. In this study, we propose a methodology to probe unique ion transport mechanisms. This strategy encompasses the use of anion exchange thin films, interdigitated electrodes, two-dimensional Infrared spectroscopy (2D IR), and atomistic Molecular Dynamics (MD) simulations. We quantify bromide ion conductivities in polynorbornene-based thin films as a function of temperature and relative humidity, harnessing electrochemical impedance spectroscopy. Intriguingly, bromide ion conductivities exhibit Arrhenius-type behavior, and, for the first time, we employ activation energy (E_a) as a diagnostic tool to discern the transition between two transport mechanisms. Our study further enriches percolation theory by integrating empirical results with MD simulations. We merged ultrafast IR and 2D IR spectroscopy with simulations to examine water dynamics and their effects on ion behavior. Quantitatively, we discovered that the transition between two mechanisms is spurred by enriched solvation environments for anions and escalated percolation of water pathways at 55% RH. This new transport mechanism is underpinned by the robust network of water molecules and a significant population of water molecules in the second solvation shell, which reduces the bridging effect and eases constraints on ions. Our findings hold considerable implications for the engineering of AEMs as proficient ion-conducting polymers.

87n Understanding Water Absorption, Percolation, and Ion Transport in an Anion Exchange Membrane

Zhongyang Wang, Ge Sun, Mrinmay Mandal, Abhishek Sharma, Chuting Deng, Kai Wang, Aaron Taggart, Alex Martinson, Paul Kohl, Juan J. de Pablo, Shrayesh Patel, Paul F. Nealey

Understanding ion transport mechanisms in an anion exchange membrane (AEM) is crucial for designing efficient electrochemical-driven processes such as fuel cells, water electrolyzers, reverse electrodialysis, and redox flow batteries. A key challenge in understanding the ion transport properties of AEMs is the lack of a methodology that captures the ion solvation environment, water network, and ion transport at different time and length scales. Here we have synthesized and fabricated polynorbornene-based anion exchange thin films as our model polymers due to their high alkaline stabilities and ionic conductivities via vapor infiltration reactions (VIRs). We customize an in situ ellipsometer to understand thin film expansion at different hydration levels and water uptakes of AEMs has been measured by dynamic water sorption. Bromide ion conductivities in polynorbornene-based thin films are measured as a function of temperature and relative humidity using electrochemical impedance spectroscopy. Bromide ion transport shows Arrhenius behaviors and activation energy (E_a) is used for the first time as an indicator for detecting the transition of transport mechanisms. Percolation theory is examined and enriched by combing experimental results and atomistic molecular dynamics simulations. We quantitatively demonstrate that the transition of a slower site hopping mechanism and a faster transport mechanism is aided by better solvation environments of anions and more percolated water pathways at a certain RH.

87o Development of High-Power and Durable MEA for Polymer Electrolyte Membrane Fuel Cells Using Structured Carbon Nanofiber Matrix

Yunseong Ji, Dae Woo Kim, Yunggun Shul

As fuel cells are being commercialized, the need to reduce the cost of platinum catalysts is becoming more pressing, as they account for approximately 40% of the final cost. Therefore, enhancing the activity and durability of platinum electrode catalysts is essential to reduce the final cost. Although various platinum-based catalysts have been developed, the performance in commercial use remains low compared to the state-of-the-art evaluated using a three-electrode system with a rotating disk electrode.

One of the main reasons for the reduced performance is the formation of a carbon-platinum-ionomer aggregate through interactions between solvents and materials and the random arrangement and drying of the aggregate to form an electrode layer.

Therefore, there is a need to develop a new electrode structure that can actively control the structure of the electrode layer to optimize the performance of the electrode. This research aims to derive optimization factors for the performance of electrodes by controlling the composition and structure of the electrode layer through active control, rather than forming a random structure of electrode materials.

A well-designed membrane electrolyte assembly (MEA) composed of electrode layers of effective materials and structure can alter the performance and durability of PEMFC. To achieve excellent electrochemical properties, active Pt nanoparticles are controlled by a nano-glue effect on a highly graphitized carbon surface. The developed MEA exhibits a notable maximum power density of 1082 mW/cm^2 at 80 °C, H_2/air, 50% RH, 1.8 atm, and low cathode loading of 0.1 mg_{Pt}/cm^2, and catalytic performance decays of only 23.18% under commercial based durability protocols, respectively.

The developed technology can be used as a platform technology that can be applied to the commercial production of fuel cells, thereby achieving all desirables for commercial applications.

87p Computational High-Throughput Study of Hydrogen Permeation through Two-Dimensional Structures for Use As Proton-Conducting Membranes

Yuting Li, Daniel Bahamon, Jaber Almarri, Nirpendra Singh, Marcelo Lozada-Hidalgo, Andre K. Geim, Lourdes Vega

Hydrogen technologies, such as fuel cells, are designed around the membrane. Aspects such as humidity and heat management, catalyst activity and poisoning, cooling systems, among others, must be controlled carefully [1]. Unfortunately, the polymeric materials used today still cannot operate at 100°C-400°C, the infamous 'materials gap' in proton conductors [2].

Hence, this work focuses on the evaluation of novel 2D-crystals to be used as proton conductive materials that could operate in conditions where typical polymeric membranes fail. The objective is to show the possibility of using such 2D-crystals as a disruptive technology addressing some of the challenges faced by hydrogen energy conversion technologies. In fact, two-dimensional crystals with angstrom-scale pores are widely considered as candidates for a next generation of molecular separation technologies aiming to provide extreme, exponentially large selectivity combined with high flow rates.

The contribution builds on the findings from Sir Andre Geim's group demonstrating that graphene, although completely impermeable to all gases at ambient conditions, is highly permeable to protons, nuclei of hydrogen atoms [3]; in addition, atomically thin micas, ~10 times thicker than graphene, have shown to be ~100 times more proton conductive than graphene [4]. Furthermore, nanoscale ripples (corrugations) in freestanding graphene membranes have been found to be catalytically active toward the H_2 dissociation [5]. This could turn completely inert materials into catalytically active, and materials that already are catalytically active in the bulk could become even more active in the 2D limit -thus reducing/removing the need to use noble metals-.

There are hundreds of two-dimensional materials known -from electron transport studies- that have not been explored from a permeability perspective. Therefore, in this contribution, several 2D-materials have been assessed, including one-atom thick structures from the graphene family, as well as hexagonal boron nitride (hBN), silicene, germanene, phosphorene, stanine, among others [6]. In addition, transition metal dichalcogenides (e.g., MoS2, MoSe2, WSe2, etc.), mxenes, and 2D covalent organic frameworks (2D-COFs) have been included in the assessment with their wider interatomic lattices and rich chemistry.

The works focuses on investigating the proton permeability of novel 2D crystals and evaluating the possible increase in the catalytic activity of 2D materials through nano-scale ripples and local curvature effects in the mechanism of H_2 dissociation reaction in 2D-materials.

Therefore, Density Functional Theory (DFT) calculations have been used as a high-throughput computational method [7] for the screening of such novel 2D-materials. This modelling tool was applied to guide the design and selection, based on two descriptors [8],[9] related to the permeability of the 2D-membranes: the electron clouds of the 2D crystals (which may impede the passage of the protons) and the energy profiles as a function of the proton distance to the basal plane of the 2D-materials (related to the proton transfer barrier), using graphene as benchmark [3]. Rigorous electronic structure calculations were performed for this endeavor, including dispersion correction of van der Waals interactions and spin-polarization by means of the VASP software. Moreover, energies for the dissociation of molecular hydrogen were performed on flat and rippled selected one-atom thick structures (including graphene). Initial "corrugated" (rippled) structures were created with a certain protrusion size by allowing the atomic structure to relax under biaxial compression. The energy barriers for the reaction pathway were performed based on climbing-image nudged elastic band (CI-NEB) calculations, explicitly including several intermediate states during the reaction process [4]. The effect of temperature and pressure into the hydrogen dissociation mechanism in these 2D-crystals was also considered.

It was found that the permeability highly depends on the density of the crystal electron clouds that fill the space between the atomic nuclei in the lattice. These clouds have low-density areas that essentially behave as 'pores' through which protons permeate. Furthermore, with respect to the nanoscale ripples, preliminary results suggest that H_2 molecules dissociate into H atoms on the ripples, leading to a finite coverage of the surface with hydrogen atoms, and with the resulting H atoms permeating via a 'flip' mechanism, similar to the previously reported for graphene [5].

The work presented here belongs to a long-term project on combining molecular modelling with experimental techniques for developing efficient proton-exchange membranes, aiming at using the fundamental understanding of the phenomena happening at the molecular level to optimize their performance at process conditions.

The funding by Khalifa University of Science and Technology under the Research and Innovation Center for Graphene and 2D-materials is greatly acknowledged.

[1] Li, R. He, J.O. Jensen, N.J. Bjerrum. Chem. Mater. 15 (2003) 4896–4915.

[2] Norby. Solid State Ion. 125 (1999) 1–11.

[3] Hu, M. Lozada-Hidalgo, F.C. Wang, A. Mishchenko, F. Schedin, ..., A.K. Geim. Nature 516 (2014) 227–230.

[4] Mogg, G.-P. Hao, S. Zhang, C. Bacaksiz, Y.-C. Zou, ..., M. Lozada-Hidalgo. Nat. Nanotechnol. 14 (2019) 962–966.

[5] Z. Sun, Q. Yang, W.J. Kuang, Y.V. Stebunov, W.Q. Xiong, ..., A.K. Geim. Nature 579 (2020) 229-232.

[6] Z. Rahman, C.W. Kwong, K. Davey, S.Z. Qiao. Energy Environ. Sci. 9 (2016) 709-728.

[7] Zhang, A. Chen, Z. Zhou. WIREs Comput. Mol. Sci. 9 (2019) e1385.

[8] J. Garcia, D. Bahamon, L.F. Vega. ACS Sustaiinable Chem. Eng. 9 (2021) 3157-3171

[9] Li, D. Bahamon, M. Sinnokrot, K. Al-Ali, G. Palmisano, L.F. Vega. J. Photochem. Photobiol. C: Photochem. Rev. 49 (2021), 100456.

392a Vermiculite/$MgSO_4$ Composite for Thermal Energy Storage

Suboohi Shervani, F Handan Tezel

Thermal energy storage using exothermic water vapor adsorption process is a novel technology to store thermal energy from the renewable energy sources and utilizing it later when it is needed. The main advantage of this technology is that it does not use any toxic chemicals, and it is pollution free. A recyclable mineral, vermiculite is a good adsorbent for thermal energy storage as it has high water sorption capacity. In this paper, water-vapour was used as the adsorbate and vermiculite/$MgSO_4$ composite was used as the adsorbent. The thermal energy storage density of the composite was determined to be 189 kWh/m^3 at an inlet relative humidity (RH) of 90 % after regeneration at 120°C using a lab scale thermal energy storage setup. Vermiculite itself had a very low energy storage density at 14 kWh/m^3 for 90%RH at the inlet after regeneration at 120°C, which showed that the composite turned out to be a good candidate in enhancing the energy storage density of vermiculite. The composite had a cost of 1.84 CAD/kWh, which showed that it is a cost effective material.

392b Characterizing Thermochemical Energy Storage Materials for Heat Decarbonization in Buildings

Akanksha Menon

The buildings sector is a significant consumer of primary energy, of which 60% is used for space conditioning. To decarbonize these thermal loads, promising solutions include solar heating and renewable electrification, but their intermittency necessitates the use of thermal energy storage (TES) to match demand and supply. Of the different TES material categories, thermochemical materials (TCMs) exhibit volumetric energy densities that are at least 3 higher than phase change materials (PCMs) and sensible heat storage, making it attractive for TES in buildings where space is limited. Furthermore, TCMs exhibit negligible heat loss (i.e., no self-discharge) as they store and release energy via reversible chemical reactions between a salt and water vapor (closed loop) or moist air (open loop). Selection of these salt hydrate TCMs requires careful consideration of multiple thermodynamic properties and material morphology that influences both heat and mass transport kinetics, leading to reaction irreversibilities and loss of storage capacity during cycling. To this end, we discuss two promising approaches to mitigate hygrothermal and structural stability challenges with TCMs: (i) encapsulation of the salt into a highly porous matrix, and (ii) formation of binary salt mixtures with complementary phase diagrams and hydration kinetics. We also show the importance of particle size on the hydration/dehydration kinetics. Overall, these composites can enhance water vapor diffusion and energy storage density, while maintaining structural stability under cycling to achieve high thermal power outputs. This approach also extends the operational window of individual salts, and it enables the addition of thermally conducting filler materials to improve the thermal conductivity of TCMs.

392c Pumped Thermal Electricity Storage: Modelling of a Reversible Recuperative sCO2 Pumped Brayton Cycle for Variable Power Operation

Zhennan Zhu, Alp Albay, Mehmet Mercangoz

Pumped Thermal Electricity Storage: Modelling of a Reversible Recuperative sCO2 Pumped Brayton Cycle for Variable Power Operation

Introduction

Decarbonization of the energy grid involves increasing use of renewable sources such as solar and wind. However, these energy sources have inherent intermittency issues. Thus, it is vital to develop highly scalable energy storage technologies to complement this increase in renewable energy generation. The focus of this work is one such technology, Pumped Thermal Electrical Energy Storage (PTES), utilizing heat pumps and heat engines to store electricity in the form of thermal energy [1,2] and more specifically, a reversible recuperated closed Brayton cycle employing supercritical CO_2 as the working fluid. In times of surplus electricity production, the process operates as a heat pump, storing incoming electrical energy in a heat storage medium as thermal energy. Conversely, during times of increased electricity demand the process is reversed, operating as a heat engine, and producing electricity. The most flexible and effective solution for the controllability of the power grid is to have such electricity storage units to operate with variable power charging and discharging rates. The current study uses Aspen HYSYS for detailed modeling of part-load operating conditions and MATLAB Simulink is used for dynamic modeling and control studies. Previous studies for modeling of part-load and off-design operation of similar systems used simplified models for compressor, turbine, and heat exchanger modeling [3,4]. In our work, these units are modeled more rigorously, with turbomachinery maps sourced from machine builders and considering temperature distributions in the heat exchangers.

The cycles operate a set of compressor-expander pairs. These have set design points at which they operate at their highest efficiency. When modelling the process, this design point is taken into consideration and the turbomachines are chosen accordingly. However, during operation the charging or discharging power may fluctuate and cause the process to deviate from this design point. It is important to understand the behavior of such systems under off-design conditions and develop control strategies to deal with these deviations. This study uses turbomachinery maps to investigate part-load operation and control strategies of such a system.

Part-load Performance Analysis

Studying the part-load characteristics is crucial to achieve high efficiencies when the power is varying [4,5]. In this study, the PTES power consumption, mass flow rate and thermal efficiency were studied. Fig. 1 shows the layout of the charge and discharge cycles of the PTES model and Fig. 2 shows the performance maps used to model the main compressor.

In summary, this study seeks to provide a detailed investigation for the full-load and part-load performance of supercritical CO_2 based recuperated Brayton cycles for Pumped Thermal Electricity Storage. Based on those results, a dynamic model is established as well as plantwide control strategies for coordinated operation with the power grid. The overall goal of this research is to achieve flexible operation capabilities with PTES systems while maintaining near-optimal performance. The steady-state and dynamic simulation results we present will provide novel insights for utilizing this new energy storage technology to its full potential.

References

[1] Mercangöz, M., Hemrle, J., Kaufmann, L., Z'Graggen, A. and Ohler, C., 2012. Electrothermal energy storage with transcritical CO2 cycles. Energy, 45(1), pp.407-415.

[2] Morandin, M., Maréchal, F., Mercangöz, M. and Buchter, F., 2012. Conceptual design of a thermo-electrical energy storage system based on heat integration of thermodynamic cycles–Part A: Methodology and base case. Energy, 45(1), pp.375-385.

[3] Frate, G.F., Paternostro, L., Ferrari, L. and Desideri, U., 2021, June. Off-Design of a Pumped Thermal Energy Storage Based on Closed Brayton Cycles. In Turbo Expo: Power for Land, Sea, and Air (Vol. 84966, p. V004T07A004). American Society of Mechanical Engineers.

[4] Yang, J., Yang, Z. and Duan, Y., 2020. Part-load performance analysis and comparison of supercritical CO2 Brayton cycles. Energy Conversion and Management, 214, p.112832.

[5] Xingyan, B., Wang, X., Wang, R., Cai, J., Tian, H. and Shu, G., 2022. Optimal selection of supercritical CO2 Brayton cycle layouts based on part-load performance. *Energy*, *256*, p.124691.

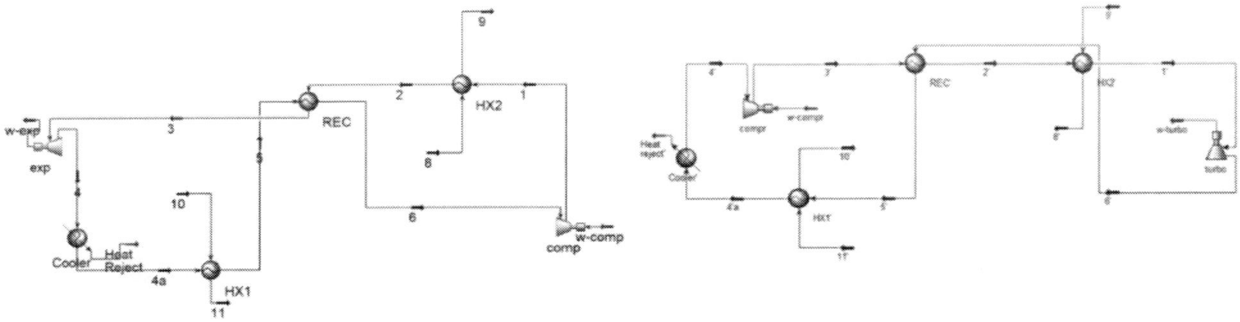

Fig. 1. The model of the charge and discharge cycles of the PTES in Aspen HYSYS

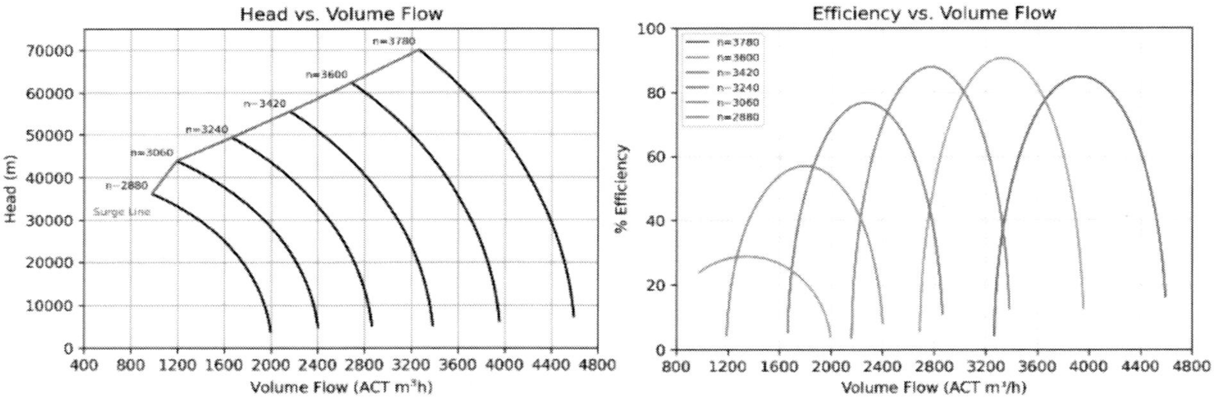

Fig. 2. Performance maps used in the studies for the main compressor

392d Application of Phase Change Materials (PCMs) in the Ceiling Fan Ventilation for the Building Thermal Energy Comfort Assessment

Seyed Mojtaba Sadrameli, Mostafa Alizadeh

Passive cooling techniques is one of the most promising technologies that adapts the principle of supplying cool air to the buildings with no or minimal electricity consumption. The system requires a thermal energy storage system packed with phase change materials (PCMs). These materials absorb the available cooling from the ambient temperature during the night and release it to the room during the day to maintain a comfortable condition. This paper presents some of the results of a study performed on the indoor thermal comfort using phase change materials on the ceiling fans. The main objective was to investigate the potential solution for overheating/ overcooling mitigation. Two laboratory scale test rooms have been designed and constructed which was controlled by linearly rising/falling temperature changing process. The performance of the system was evaluated experimentally by comparing the performance of test room with PCM and a similar room without PCM. The experiments were conducted in an artificial climate inside a laboratory environment chamber which was controlled by harmonic and linearly rising/falling temperature changing processes. Experimental design techniques based on 5 level RSM CCD method to quantify the potential effects of using PCM panels and to determine individual/interactive effects of parameters on thermal discomfort index, PPD. The results indicate that minimum discomfort level can be achieved when inlet air temperature and humidity, fan rotating speed and PCM slab height and thickness were set to 29.48%, 115 rpm, 31 cm and 2.6 cm, respectively. PPD was obtained 4.1%in the optimal condition. In summertime conditions the results prove that the PCM used in one of the rooms lead to an overheating reduction of 13.83% representing PCM efficiency of 56%. Under the winter conditions during two months, the results show that the proposed hybrid system can reduce discomfort level around 2.61% corresponding a PCM efficiency of 35,5%. Based on the results the proposed system has a great potential to shift cooling/heating energy demand away from peak hours.

392e Mathematical Model of Energy Storage By High Temperature Microencapsulated Phase Change Material

Sk. Moumita Manjur, Sidney Lin

Enormous amount of heat released by industrial processes can be reused to improve energy efficiency as well as to reduce heat pollution for a sustainable and greener environment. Latent heat storage using phase change material (PCM) is a promising technique to capture and recycle industrial waste heat. Use of high temperature (m.p. >300°C) PCM in a large scale is still under development. A mathematical simulation can help understand the heat transfer behavior and energy storage efficiency to optimize the heat storage system design and operation. Microencapsulated phase change materials (MEPCMs) consisting of phase change materials packed inside a resilient shell (e.g., polymer, alumina, sodium silicate or steel) have advantages of isolation from the external environment, increasing the heat exchange area and preventing leakage. In addition, MEPCMs can be transported between remote site and heat source safely and economically. This work studies the heat transfer behavior of high temperature MEPCMs during charging (heating) and discharging (cooling) processes.

A finite element analysis model of heat transfer in a phase change material pushing towards phase transition has been developed using COMSOL Multiphysics®. In this model, temperatures at different time, location, and PCM phase (solid or liquid) inside the MEPCM of an alumina shell have been calculated. Temperatures of MEPCM at solid and liquid phases were calculated using Fourier's law of heat conduction. Temperature during solid/liquid phase transition (melting point) was calculated numerically using the general concept of first order phase transition by absorbing/releasing fixed amount of heat at a constant temperature. Temperature and composition dependent thermophysical and thermodynamic properties of the PCM and alumina reported in literatures were used in this model. Three different heating/cooling conditions were investigated in this work: i) the outermost surface of the MEPCM is maintained at a constant temperature; ii) a constant heat flux is applied on the MEPCM shell surface, and iii) a convection heat transfer condition is used for heat exchange between environment and MEPCMs. The calculated results are analyzed to study the phase change phenomenon of MEPCM which reflects the overall performance of MEPCM heat storage during charging and discharging process.

480a Advanced Catalysts for Hydrogen Fuel Cell Technologies

Gang Wu

Due to the overwhelming use of fossil fuels in past decades, clean and efficient energy storage and conversion via sustainable electrochemical reactions associated with hydrogen, oxygen, and water have attracted substantial attention for energy and environmental sustainability. Among compelling energy technologies, hydrogen proton exchange membrane fuel cells (PEMFCs) are a promising zero-emission power source for transportation to mitigate environmental pollution and reduce fossil-fuel dependence. Relative to traditional family cars, PEMFCs are highly desirable for heavy-duty vehicles (HDVs) requiring high efficiency, long driving distances, and fast refueling. However, current materials, including catalysts, membranes, and ionomers, cannot meet the challenging targets of high-efficiency, low-cost, and long-term durability of up to 25,000 hours (or one million miles) for HDVs. Developing high-performance cathode catalysts is crucial for significantly improving performance and reducing the costs of PEMFCs for HDV applications. Our research group at SUNY-Buffalo has been well supported by U.S. DOE in the past decade, aiming to address materials issues through designing and scaling up innovative and highly efficient catalysts and electrodes.

In this talk, I discuss recent understanding, progress, achievement, and perspective on developing low-cost and high-performance air cathode catalysts via two types of technique solutions, i.e., platinum group metal (PGM)-free atomically dispersed metal-nitrogen-carbon catalysts and ordered low-PGM intermetallic PtCo catalysts for PEMFCs.

480b From PGM to Low PGM to PGM-Free Catalyst Layers for AEM Electrolyzers

Mahmoud Amirsalehi, Noor ul Hassan, Hari Gopi, William Mustain

Anion exchange membrane water electrolyzers (AEMELs) have been widely touted as the holy grail for renewable water electrolysis – combining the advantages of PEM electrolyzers (high current density, high pressure operation, etc.) with the advantages of traditional alkaline electrolyzers (e.g. the use of low cost components). To date, the potential for AEMELs to supplant these two incumbent technologies has not been realized due to a number of issues. One of the most pressing issues is the activity and durability of catalysts, necessitating the use of high loadings of platinum group metal (PGM) catalysts at both the anode and cathode. This talk will discuss the optimization of PGM-based electrodes to improve performance as well as the transition of the learning involved in that development to low-PGM and PGM-free catalyst layers. The primary focus of the talk will be on performance and durability as PGM-based materials are replaced at the anode and cathode. The talk will conclude by defining pathways for future research.

480c Innovations in Proton Exchange Membrane Fuel Cell Cathode Materials and Architectures for High Power Density, and Durability

Shawn Litster

This talk will highlight our group's recent research on advancing PGM cathode performance and durability, including the development of novel diagnostics, imaging methods, and simulation tools. The talk will highlight three key areas of innovation for PGM cathodes: 1) highly accessible catalyst supports, 2) high oxygen permeability ionomer for electrodes, and 3) ionomer-free cathodes. First, the talk will highlight the analysis of transport, reaction, and degradation in highly accessible catalysts, including image-based modeling of the coupled transport and reaction in catalyst aggregates that yields new insights into transport resistances in the support as well as Pt particle size effects due to electrokinetic phenomena. The analysis also includes evaluating the redistribution of Pt at the aggregate-scale during voltage cycling with high surface area carbon supports. The talk will further highlight the use of the nano-scale resolution X-ray CT (nano-CT) at Carnegie Mellon for imaging the redistribution of Pt at the MEA scale. Second, the talk will cover our progress on integrating high oxygen permeability ionomer (HOPI) into cathodes for higher performance and durability as part of the Carnegie Mellon-led project in the DOE's Million Mile Fuel Cell Truck (M2FCT) program. The talk will present our studies on optimizing HOPI inks and coatings, as well as highlighting our experimental and modeling results that unveil the mechanisms behind the higher apparent mass activity and improved durability with HOPI. We will also introduce our approach to evaluating the durability of new electrode ionomers via a simple electrode-specific AST. Finally, the talk will highlight our recent work on ionomer-free cathodes that use novel catalysts to eliminate the need for ionomer and alleviate its negative impacts of reduced ORR activity and high oxygen mass transport.

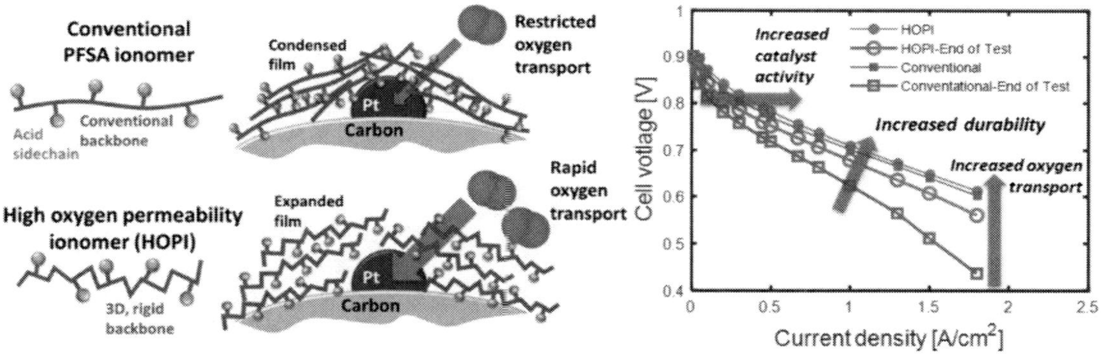

480d Controlling Water at Catalyst/Ionomer Interface with Molecular Additives

Joshua Snyder

The growth and viability of deep decarbonization through the hydrogen economy is dependent upon scientific advances that will address the existing political, regulatory, and technological barriers that currently hinder electrochemical energy technologies. The strong correlation between device-level performance and catalytic electrode activity/durability highlights the criticality of identifying the limiting processes and developing strategies to address these limitations. For proton exchange membrane fuel cell and water electrolyzers (PEMFC and PEMWE), many of the existing limitations can be traced to the interface between catalyst materials and ion conducting electrolytes, be they aqueous-based or solid electrolytes. The catalyst/ionomer interface is plagued by negative impacts to both reaction kinetics and reactant transport. Mitigating these impacts while maintaining sufficient ionic transport and contact in device catalyst layers is challenging. In this talk, we will present our work with molecular interfacial modifying species and their role in alleviating many of the detrimental impacts of the ionomer/catalyst interface. Additionally, we will highlight the role of water in defining catalyst performance at these interfaces and how we can optimize the interaction of solvating water with the catalyst surface to improve reaction kinetics.

480e Novel Porous Transport Layers for Polymer Electrolyte Membrane Water Electrolyzers

Siddharth Komini Babu, Tanvir Arman, Abdurrahman Yilmaz, Jacob LaManna, Daniel S. Hussey, Jacob Spendelow

A significant barrier to the growth of the hydrogen economy is reducing the cost of hydrogen.[1] Hence, the US DOE's hydrogen earth shot initiative aims to achieve the cost target of $1 per 1kg of hydrogen in 1 decade.[2] Polymer electrolyte membrane water electrolyzer (PEMWE) is a vital electrochemical technology for generating clean hydrogen. PEMWEs still requires a significant reduction in capital and operational cost to meet the cost targets. Thrifting iridium to reduce the cost of PEMWE affects the efficiency due to poor contact with the porous transport layers (PTLs) and decreased catalyst utilization.[3] Moreover, PTLs morphology also controls the water transport to the catalyst layer while simultaneously removing the generated O_2 bubbles, affecting mass transport.[4] Optimal PTL morphology can improve the interface and the mass transport, increasing the efficiency of PEMWE and thereby reducing the cost.[5] This work presents the development of novel PTLs architectures for PEMWE to improve performance. Hierarchical porous PTLs are fabricated by adding microporous layers (MPLs) through the deposition of conductive powder with a binder by decal transfer. The wettability of the MPL is controlled by changing the loading and the type of binder. Electrochemical evaluation of the PTLs, including the polarization curve and contact resistance, will be presented to elucidate the effect of the PTLs on the performance. In addition, neutron imaging of an operating PEMWE will be presented to study the impact of the hierarchical PTLs on water management.

Acknowledgment

This research is supported by the U.S. Department of Energy (DOE) Hydrogen and Fuel Cell Technologies Office through the H2NEW consortium.

References

1. A. Majumdar, J. M. Deutch, R. S. Prasher and T. P. Griffin, *Joule*, 2021, **5**, 1905-1908.
2. B. S. Pivovar, M. F. Ruth, D. J. Myers and H. N. Dinh, *The Electrochemical Society Interface*, 2021, **30**, 61-66.
3. T. Schuler, J. M. Ciccone, B. Krentscher, F. Marone, C. Peter, T. J. Schmidt and F. N. Buchi, *Advanced Energy Materials*, 2020, **10**, 1903216.
4. J. K. Lee, C. Lee, K. F. Fahy, B. Zhao, J. M. LaManna, E. Baltic, D. L. Jacobson, D. S. Hussey and A. Bazylak, *Cell Reports Physical Science*, 2020, **1**, 100147.
5. X. Peng, P. Satjaritanun, Z. Taie, L. Wiles, A. Keane, C. Capuano, I. V. Zenyuk and N. Danilovic, *Adv Sci (Weinh)*, 2021, **8**, e2102950.

480g High Throughput Screening of MEA Performance and Durability Using Fuel Cell Short-Stacks

Tim Van Cleve, K.C. Neyerlin

The transportation sector currently accounts for a large portion of emissions of both greenhouse gasses, which contribute to climate change, and particulates, which have many adverse health effects. Decarbonization efforts will rely on the rapid development and deployment of electric vehicles. Hydrogen fuel cells have received considerable interest as a cleaner replacement for diesel engines for long distance transportation, as fuel cells can generate more power than batteries and are not range limited. To outcompete diesel engines, fuel cell systems must operate with high fuel efficiency and durability to lower the total cost of ownership to a competitive level. With these ambitious targets, we need to rapidly integrate promising materials in membrane electrode assemblies (MEAs) and assess their performance and durability to ensure their competitiveness.

The development of high-performance and durable MEAs is challenging due to the difficult optimization of many processing and fabrication variables (e.g. ink formulation and mixing, deposition method, solvent ratio, hot-pressing, conditioning, etc) that can significantly impact overall performance and durability. Unfortunately, this process is highly empirical therefore requires substantial time and resources when incorporating a novel component. There's a clear need to accelerate the ability to assess and identify key design parameters to advance the development of hydrogen fuel cell technology.

To address the limitations in testing capacity, we utilize short stacks to rapidly screen MEAs up to 10 times faster than single cell testing. Levering our experience in electrochemical diagnostics development, we have recently developed galvanostatic techniques to monitor critical MEA properties (like hydrogen crossover, electrochemical surface area, and ohmic resistance) of individual cells in the stack. Coupling electrochemical characterization with high throughput performance screening, our goal is to fast-track the discovery of structure-performance relationships essential to durable fuel cell development. This presentation will highlight these new capabilities and their applications when assessing the degradation of MEAs over the course of accelerated stress tests.

Digested Extracellular DNA Shortens the Anodic Startup of Microbial Electrolysis Cell

Jiaman Wang

While the multiple functions of extracellular DNA (exDNA) in biofilm formation and electron transfer have been extensively studied in pure culture, its role in mixed anodic biofilm was still unknown. In this study, we employed DNase I enzyme to digest exDNA. thereby investigating its role in anodic biofilm formation based on the performance of four microbial electrolysis cells (MECs) groups with different DNase I enzyme concentration (0, 0.05, 0.1, 0.5 mg/mL). The responding time to reach 60 % maximum current of treatment group with DNase I enzyme has been significantly reduced to 83 %–86 % of the blank group (t-test, $p < 0.01$), indicating the exDNA digestion could promote the biofilm formation at the early stage. The anodic coulombic efficiency was enhanced by 10.74– 54.42 % in treatment group (t-test, $p < 0.05$), which could be ascribed to the higher absolute abundance of exoelectrogens. The lower relative abundance of exoelectrogens indicated the DNase I enzyme addition was beneficial for the enrichment of extensive species rather than exoelectrogens. As the DNase I enzyme augments the fluorescence signal of exDNA distribution in the small molecular weight region, implying the short chain exDNA could contribute to the biomass enhancement via boosting the most species enrichment. Furthermore, the exDNA alteration improved the complexity of microbial network. Our findings provide a new insight into the role of exDNA in the extracellular matrix of anodic biofilms.

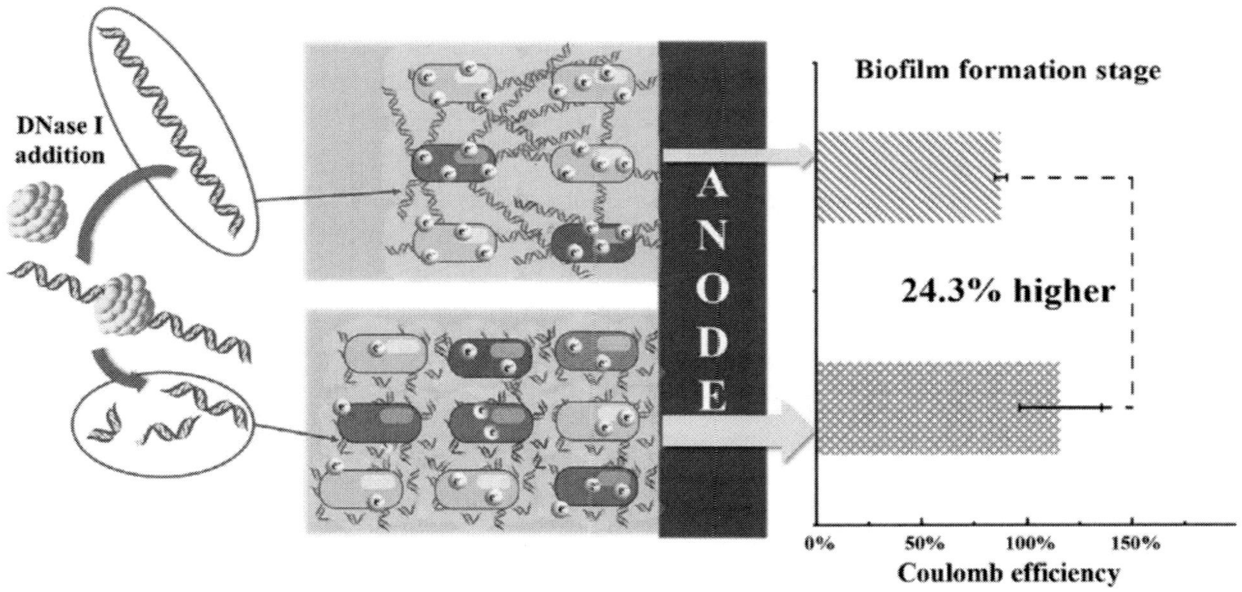

232a Ion Transport in Li^+ Polyelectrolytes and Concentrated Binary Salt Electrolytes

Helen Bergstrom, Kara Fong, Bryan McCloskey

Conventional Li-ion battery electrolytes have been designed to optimize numerous desirable properties, including interfacial and thermal stability, conductivity, and low flammability. However, all Li^+-bearing electrolytes still possess low Li^+ transference (t_+) numbers, where current passed through them is primarily carried by the counteranion, resulting in large concentration gradients that limit battery performance, particularly at high discharge and charging rates. The development of high t_+ electrolytes—those in which most (or all) current is carried by the Li^+ ion—could enable safer battery cycling, faster charging rates, and thicker, more energy-dense cathode designs in Li-ion batteries.

In recent years, both polyelectrolyte solutions (charged polymers dissolved in solvent) and highly concentrated binary salt electrolytes (or so-called solvent-in-salt electrolytes) have been reported to provide high Li^+ transference numbers, as measured using the popular Bruce-Vincent method.[1] In this presentation, we will discuss the complete transport characterization of model systems in both electrolyte classes. We show that, for modest-to-highly conductive (>1 mS/cm) electrolytes, voltage polarization methods in Li|Li symmetric cells are hindered by large Li interfacial resistances,[2] thereby limiting the utility of the Bruce-Vincent method that relies on this measurement. We therefore implement electrophoretic nuclear magnetic resonance to direct quantify ion velocities through an applied electric field, allowing ion transference numbers to be measured without interfacial artefacts.[3] Ultimately, in polyelectrolytes and solvent-in-salt systems, ion correlations result in limited t_+ well below those reported in the literature. Furthermore, for polyelectrolytes with roughly 10 repeat units and greater, we observe that the average velocity of Li ions in an electric field is in the direction opposite than expected, evidence of a negative t_+ due to correlated motion through Li ion condensation on the polyelectrolyte chain. While this result is a negative outcome in our pursuit of high t_+ electrolytes, we anticipate many of the capabilities and insights gleaned throughout our studies to be of general use to those studying novel, concentrated electrolyte solutions where strong, non-ideal ion-ion correlations are observed.

References

1. Evans, J.; Vincent, C. A.; Bruce, P. G. Electrochemical measurement of transference numbers in polymer electrolytes. *Polymer* **1987**, *28*, 2324-2328. http://dx.doi.org/10.1016/0032-3861(87)90394-6
2. Bergstrom, H. K.; Fong, K. D.; McCloskey, B. D. Interfacial effects on transport coefficient measurements in Li-ion battery electrolytes. *Journal of the Electrochemical Society* **2021**, *168*, 060543. 10.1149/1945-7111/ac0994
3. Bergstrom, H. K.; Fong, K. D.; Halat, D. M.; Karouta, C. A.; Celik, H. C.; Reimer, J. A.; McCloskey, B. D. Ion correlation and negative lithium transference in polyelectrolyte solutions. *Chemical Science* **2023**, *14*, 6546-6557. 10.1039/D3SC01224G

232b Reaction Mechanisms in $LiPF_6$ carbonate Electrolytes Modified By Phosphorous Pentoxide and Their Effect on Lithium Metal Interface

Juchen Guo

Rechargeable lithium (Li) metal batteries are the most promising electrochemical energy storage technology to break through the energy density barrier. However, the reductive decomposition and chemical instability of the liquid electrolytes against Li metal result to unstable solid-electrolyte interphase (SEI) formation and porous Li deposition with low coulombic efficiency and severe volume expansion. We demonstrate that the commercial Li-ion battery electrolyte composed of 1 M lithium hexafluorophosphate ($LiPF_6$) in ethylene carbonate and diethyl carbonate (EC/DEC 50/50 vol. ratio) modified by phosphorous pentoxide (P_2O_5) can significantly improve the cycle stability of Li metal batteries with realistic parameters in 0.4 Ah pouch cells composed of $LiNi_{0.6}Mn_{0.2}Co_{0.2}O_2$ (NMC622) cathode (3 mAh cm^{-2}) and Li anode (50 mm thick) with a lean electrolyte/capacity ratio of 3 g/Ah. In this presentation, we discuss a number of intertwined and competing reaction mechanisms between P_2O_5 and the electrolyte, revealed by nuclear magnetic resonance (NMR) spectroscopy. We also discuss the methods to control the mechanisms to produce the desirable SEI on Li metal interface with the electrolyte.

232c Submillimeter Bundled Microtubular Flow Battery Cells with Ultrahigh Volumetric Power Density

Alexandros Filippas, Nian Liu

Flow batteries are a promising energy storage solution. However, the footprint and capital cost need further reduction for flow batteries to be commercially viable. The flow cell, where electron exchange takes place, is a central component of flow batteries. Improving the volumetric power density of the flow cell (W/Lcell) can reduce the size and cost of flow batteries. While significant progress has been made on flow battery redox, electrode, and membrane materials to improve energy density and durability, conventional flow batteries based on the planar cell configuration exhibit a large cell size with multiple bulky accessories such as flow distributors, resulting in low volumetric power density. In this talk, I will introduce a submillimeter bundled microtubular (SBMT) flow battery cell configuration that significantly improves volumetric power density by reducing the membrane-to-membrane distance by almost 100 times and eliminating the bulky flow distributors completely. Using zinc–iodide chemistry as a demonstration, our SBMT cell shows peak charge and discharge power densities of 1,322 W/Lcell and 306.1 W/Lcell, respectively, compared with average charge and discharge power densities of <60 W/Lcell and 45 W/Lcell, respectively, of conventional planar flow battery cells. The battery cycled for more than 220 h corresponding to >2,500 cycles at off-peak conditions. Furthermore, the SBMT cell is compatible with zinc–bromide, quinone–bromide, and all-vanadium chemistries. The SBMT flow cell represents a device-level innovation to enhance the volumetric power of flow batteries and potentially reduce the size and cost of the cells and the entire flow battery.

232d Electrolyte Design for Beyond Li-Ion Batteries

Chibueze Amanchukwu

Batteries with high energy densities (energy stored per mass or volume) are required to electrify transportation and the grid. Lithium metal batteries can double the energy density of current lithium-ion batteries by replacing the graphite anode with a lithium metal anode. However, lithium metal is highly reactive, electrodeposits in a high surface area dendritic morphology, and continuously decomposes the electrolyte. Solid state electrolytes are promising because of nonvolatility and safety, but suffer from high electrode/electrolyte interfacial resistance. In contrast, liquid electrolytes have high ionic conductivities and low interfacial resistance but are volatile and flammable. Here, we develop solvent-free electrolytes that maintain the benefits of nonvolatile solids while enabling an improved electrode/electrolyte interface. We probe the influence of these electrolytes on lithium metal deposition and stripping, interfacial composition, and long term lithium metal cycling. Our work illustrates the importance of new electrolyte classes for next generation lithium-based batteries.

232e Ester-Based Electrolytes for Na-Ion Battery

Yunan Qin

Sodium-ion batteries (SIBs) emerge as an alternative or complement to the prevailing lithium-ion batteries (LIBs), and it have attracted a decade of intensive research efforts. In the research of electrolyte design, non-aqueous carbonate solvents (the combination of cyclic and linear carbonate) have been the main choice for the SIBs. However, carbonated-based electrolyte has limitations in ionic transport and melting point, thus limit fast charge and low temperature performance. Esters with low viscosity and low melting point are usually used as co-solvent in carbonated-based electrolyte to enhance the fast charge and low temperature performance. However, the use of linear or cyclic esters as major solvent in SIBs have not been explored. There are several studies use linear esters as sole electrolyte solvents in LIBs, but the differences between sodium and lithium batterie in terms of their chemistry/electrochemistry properties open opportunities to apply esters in SIBs. In this project, we investigate the physical and electrochemical performance of a series of esters in the context of SIBs electrolytes, figure out the relationship between the chemical structure of solvent with the Li ion solvent shell, and provide strategies to optimize the electrolyte formulation.

232f Enabling Sodium Batteries with Wide Ranging Temperature Operability

Weiyang Li

Lithium-ion batteries have been extensively employed in portable electronics and electric vehicles because of their high energy/power density. However, they inevitably suffer from severe energy/power losses in cold environments, especially when temperatures drop below −20 °C. Such poor low-temperature performance limits their applications for aeronautics/space missions, polar expeditions and many military and civil facilities in cold regions, in which a battery operating temperature below −40 °C is typically required. Therefore, improving the battery performance (energy density, rate capability and lifetime) at low temperatures with minimum need for ancillary thermal systems requires exploiting new battery chemistries beyond the lithium-ion batteries. Searching for a system with high electrochemical activity beyond the lithium technologies may address the challenges associated with low-temperature operation. Sodium has a lower first ionization energy than lithium, which may facilitate the electrochemical reactions in cold environments. In this talk, I will present our recent efforts to realize diverse sodium battery systems with a widely operating temperature range down to −80°C through innovative sodium electrochemistry and electrolyte engineering.

232g Electrochemical and Physicochemical Properties of Highly Concentrated Acetate-Based Electrolytes for High-Voltage Aqueous Zinc Batteries

Debayon Dutta, Damon E. Turney, Surabh KT, Timothy N. Lambert, Robert Messinger, Sanjoy Banerjee

Zinc-based aqueous battery systems are attractive for grid-storage applications due to the inherent electrolyte safety, material non-toxicity, low cost, high volumetric capacity of zinc and the low polarizability of the zinc anode in aqueous solutions. Despite these key features, their relatively low energy density is limited by the working potential range for aqueous electrolytes. In addition, zinc hosts a slew of problems such as inhomogeneous material distribution upon cycling and the formation of inactive zinc species and dendrites, all of which contribute to capacity loss and ultimately, cell failure. Water-in-salt electrolytes (WiSE) have been found to be an effective strategy to expand the electrochemical window of aqueous electrolytes for high energy density aqueous-based batteries, including zinc batteries. These classes of electrolytes have also been recognized in literature to provide uniform and reversible metallic deposition. There is a growing amount of literature focusing on highly concentrated acetate-based electrolytes as low-cost and non-toxic WiSE for a multitude of di- and monovalent battery systems. However, work is scarce on the nature of zinc electrochemistry, morphology and reversibility in these acetate-based concentrated electrolytes at realistic current densities and higher zinc utilization. Understanding the electrochemical behavior of zinc in acetate-based WiSE will significantly aid in the development of a high-voltage zinc aqueous battery.

In this work, the physicochemical and electrochemical properties of bi-salt potassium-zinc acetate-based WiSE were studied and compared with those of traditional alkaline potassium hydroxide (KOH) electrolytes, such as the density, viscosity, equilibrium vapor pressure, ionic conductivity, and electrochemical stability window. The impact of the high salt concentration on the electrochemical working window of the electrolytes was measured using constant-current techniques, with the highest electrochemical window observed for 27 m potassium acetate solution at 3.13 V. Zinc symmetric cells were cycled for the first time at current densities at industry-relevant current densities of >10 mA/cm^2 to examine zinc reversibility with respect to zinc depth-of-discharge (DOD), which were compared with zinc cycling in traditional 25% potassium hydroxide (KOH) electrolytes. The zinc surface morphology and roughness in these WiSE and KOH solutions were also examined using profilometry studies on cycled zinc foil anodes. Rotating disk electrode (RDE) experiments were used to quantify hydrogen evolution at 100 and 300 mV overpotentials (vs. Zn/Zn^{II}) to give us insights on the competition between mass transfer and electrochemical kinetics. Single-pulse ^1H and ^{13}C nuclear magnetic resonance (NMR) spectroscopy were conducted in conjugation with molecular dynamic simulations (MD) to elucidate solvation structures as a function of potassium and zinc acetate concentration, while pulsed field gradient (PFG)-NMR was performed to quantify the self-diffusion coefficients of water and acetate. These molecular and transport properties were used to better understand and explain the physicochemical and electrochemical properties of the WiSE electrolytes. Overall, these studies lay the scientific groundwork for understanding the use of concentrated acetate-based electrolytes with zinc electrodes and for the technological pursuit of a practical high-voltage aqueous zinc battery.

This work was supported by the U.S. Department of Energy Office of Electricity. Dr. Imre Gyuk, Director of Energy Storage Research at the U.S. Department of Energy Office of Electricity, is thanked for his financial support of this project. This work was performed, in part, at the Center for Integrated Nanotechnologies, an Office of Science User Facility operated for the U.S. Department of

Energy (DOE) Office of Science. This work was also supported by the United States Nuclear Regulatory Commission under grant number NRC- HQ-60-17-G-0030. The views expressed in this article do not necessarily represent the views of the U.S. Department of Energy or the United States Government. This article has been co-authored by an employee of National Technology & Engineering Solutions of Sandia, LLC under Contract No. DE-NA0003525 with the U.S. Department of Energy (DOE). The employee owns all right, title and interest in their contribution to the article and is solely responsible for its contents. The United States Government retains and the publisher, by accepting the article for publication, acknowledges that the United States Government retains a non-exclusive, paid-up, irrevocable, world-wide license to publish or reproduce the published form of this article or allow others to do so, for United States Government purposes. The DOE will provide public access to these results of federally sponsored research in accordance with the DOE Public Access Plan https://www.energy.gov/downloads/doe-public-access-plan.

407a Synthesis of Metal Oxide Coatings on Silicon Nanoparticles Via Condensed Layer Deposition for Enhanced Lithium-Ion Battery Anodes

Yangchuan Xing

Silicon-based lithium-ion batteries (LIBs) are considered by many to be the future of LIBs due to silicon's high theoretical capacity of ~ 4200 mAh/g. However, upon lithiation, silicon expands by approximately 300% which can cause destruction of an LIB anode thus preventing commercial applications. As such, this project aimed to apply the thin coating technique known as the Condensed Layer Deposition (CLD) method to synthesize conductive, uniform metal oxide nanocoatings on the surface of silicon nanoparticles (SiNPs) to create more cycle stable silicon-based anodes in LIBs. Transmission electron microscopy (TEM) studies have shown that uniform coatings on the order of 2-5nm in thickness can be synthesized. Process optimization has determined that the condensation of a thin water film on the SiNP substrate is the rate determining step. High-resolution TEM and energy dispersive X-ray spectroscopy have also confirmed the presence of a uniform metal oxide coating on the surface of the SiNPs. Additional characterization of these coated SiNPs will be performed via powder X-ray diffraction and thermogravimetric analysis. The changes in cycle stability of anodes constructed from pure SiNPs versus those constructed from metal oxide-coated SiNPs are tested using coin-type half-cell electrochemical testing.

407b Surface-Engineered Metal Anodes for Aqueous Batteries

Yang Yang

Safety and durability concerns caused by surface and interface instabilities of high-surface-activity energy materials are challenging modern electrochemical energy conversion and storage systems. In this presentation, I will present our most recent representative works on the surface engineering of functional materials for aqueous batteries. In this work, we designed an *in-situ* visualized protocol that can exactly mimic the actual electrochemical environments used in aqueous batteries. Using this novel protocol, we observed the dynamic processes of metal plating/stripping on the alloys in aqueous batteries. Comprehensive and systematic studies based on experimental, theoretical, spectroscopic, and microscopic approaches prove that absolute reversibility can be achieved by using the proposed 3D alloys because of the significantly improved Zn diffusion and successfully suppressed dendrite growth. As a proof-of-concept, the novel alloy structures deliver unprecedented stability with nearly 100% Coulombic efficiency over thousands of cycles even under harsh electrochemical conditions, including testing in seawater-based aqueous electrolytes and using a high current density of 80 mA cm^{-2}.

407c Imagining Nitrogen Fixation at Lithium Interphases Via Cryo-EM

Yuzhang Li

Since early efforts by PI Li to adapt cryo-electron microscopy (cryo-EM) towards battery research, significant breakthroughs have been made in our fundamental understanding of how batteries operate and fail. **Now, our present work further advances cryo-EM towards the field of electrocatalysis for the first time, providing new findings that have eluded past studies.** Combined with a multiscale approach, our cryo-EM experiments reveal the precise role of metallic lithium and its solid electrolyte interphase (SEI) in the electrosynthesis of ammonia. Contrary to conventional wisdom, we discover that the SEI layer prevents nitrogen fixation and must be disrupted for catalytic activity. This surprising finding completely revises our previous understanding of lithium-mediated ammonia synthesis and provides new insights for designing electrochemical systems to decarbonize future fertilizer production (**Fig. 1**).

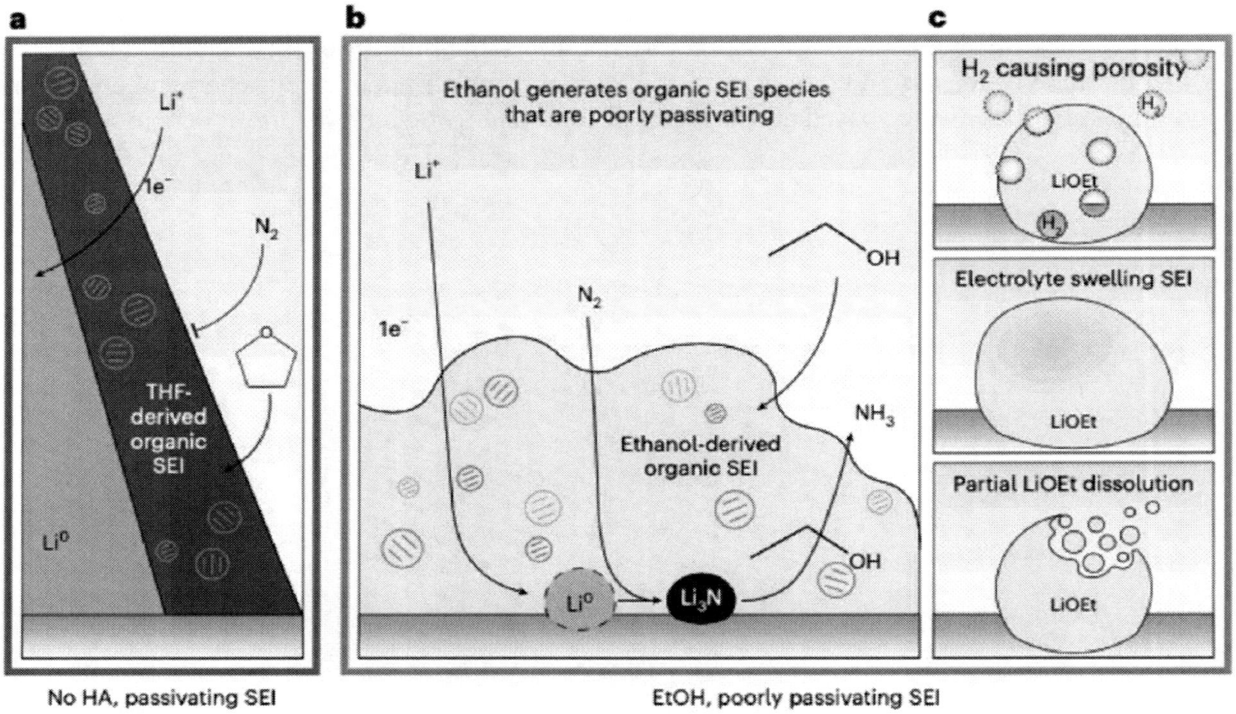

407d Partially Carbonized Polymer Binder with Polymer Dots for Silicon Anodes in Lithium-Ion Batteries

Zhuoying Wu

Large-scale applications of conventional conductive binders for Si anodes in lithium-ion batteries are challenging to accomplish due to their complex synthesis steps and high cost. Herein, a carbonized polymer dots-assisted polyvinyl alcohol-chitosan (PVA-CS-CPDs) binder is developed through a simple and low-cost hydrothermal method. Through rational design, the PVA-CS-CPDs binder retains rich polar groups while forming conjugated structures. The conjugated structure endows the PVA-CS-CPDs with high electronic conductivity, and the retained polar groups maintain strong binding strength. The proposed water-soluble binding system acts as both binder and conductive additive, enabling stable cycling for high-Si-content (90 wt %) anodes without any other conductive additives. The silicon anode consisting of 90 wt% Si and 10 wt% PVA-CS-CPDs binder shows a high reversible capacity of more than 2000 mA h g^{-1} after 200 cycles at a current density of 0.2 C and maintains an area capacity of 3 mAh cm^{-2} for more than 100 cycles with high silicon loading of 3 mg cm^{-2}. Considering these advantages, CPDs-assisted polymer binders provide a new strategy for further increasing the content of Si or other active materials in electrodes.

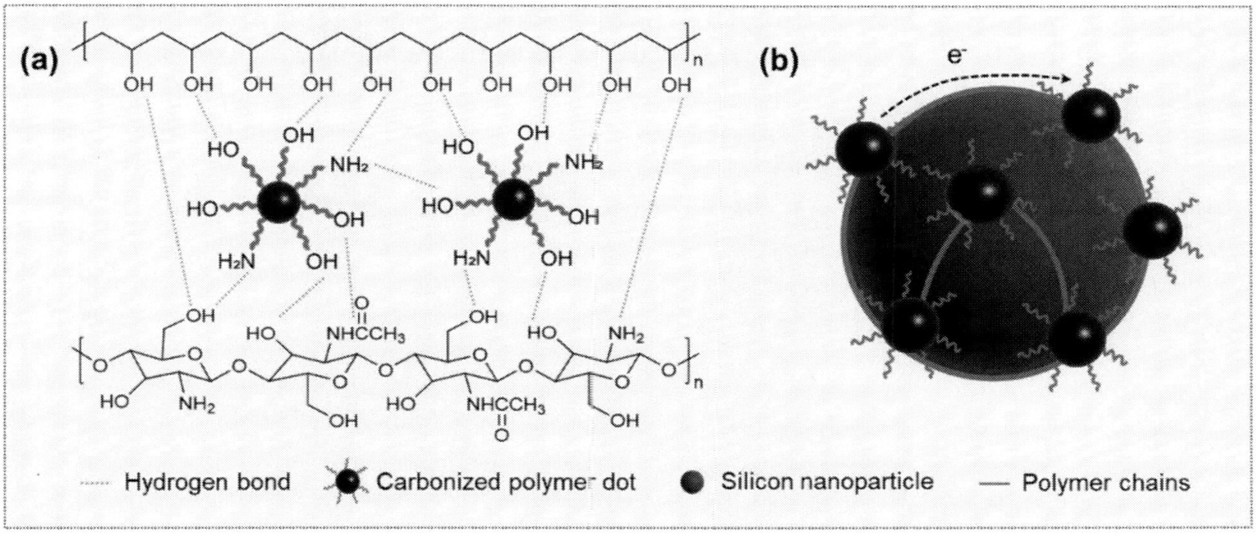

407e Effects of Surface Additives and Electrolyte pH on Potential of Zero Charge and Electrochemical Hydrogen Kinetics

Maureen Tang

It has long been recognized that both HER and HOR are multiple orders of magnitude slower in alkaline than acidic conditions for most (but not all) metals, most notably Pt. Despite decades of study, the community still lacks a unified explanation for this observation. Neither experiment nor simulations have been able to provide evidence that pH causes the intrinsic hydrogen binding energy (HBE) to change. Electric field strength has been hypothesized to be responsible for the sluggish kinetics of HER/HOR at high pH. If solvent molecules can move most freely when the field strength is minimized, the potential of zero charge should correlate to faster water rearrangement, higher frequency of the Arrhenius prefactor, and lower transition state barriers. Our previous work has used single-crystal voltammetry to study the importance of the interfacial electric field strength on hydrogen reaction kinetics. By correlating the pzfc with reaction kinetics, varying the electrolyte ionic strength at constant pH, and probing solvent dynamics via kinetic isotope effects (KIEs), we found that pzfc seems to correlate with HER/HOR kinetics, but also suggest that the effect more likely originates from activation barriers to the reaction elementary steps than from electric field effects on solvent dynamics. Here, we further explore the role of surface additives such as caffeine on solvent dynamics and overall hydrogen electrocatalysis. Our work elucidates the role of "beyond adsorption" phenomena on importance for activity and selectivity in heterogeneous electrochemical transformations.

407f Lithium-Ion Battery Cathode-Electrolyte Interface

Hongmei Luo

The global energy crisis coupled with the consumption of fossil fuels and the associated environmental issues has stimulated extensive interest in searching for clean and sustainable energy storage and conversion systems. Rechargeable lithium-ion batteries (LIBs) are widely used in mobile devices, hybrid, plug-in hybrid, and electric vehicles. Electrode-electrolyte interface has a strong impact on battery capacity decrease and long-term cycling stability. In order to improve fundamental understanding of the involved phenomena to accelerate the rational design of cathode materials and stable interface, we will discuss the cathode degradation mechanism and the processes occurring at the cathode-electrolyte interfaces. Using $LiMnO_4$ and $Li(Ni,Co,Mn)O_2$ (NCM) as cathode materials, we will focus on electrochemical characterization, cation mixing, oxygen evolution, and thin film deposition with controlled crystal structures and composition. We will further introduce our work employing the scanning electrochemical microscope to study LIBs degradation and cathode-electrolyte interface in situ and in real-time.

407g Understanding Organic Components of Solid Electrolyte Interphase on the Lithium Metal Anode

Jeffrey Lopez

Lithium ion batteries have become the dominant form of energy storage used in consumer electronics and, recently, electric vehicles. However, high costs have prevented widespread deployment of lithium ion batteries for applications other than portable electronics, and the safety issues associated with liquid organic electrolytes remain to be addressed. In order to enable the greater utilization of electric vehicles, allow for grid scale energy storage, and meet the demands of new electronic applications, new materials for high energy density batteries must be developed. High capacity electrode materials like lithium metal have the potential to facilitate these technologies, but lithium metal electrodes are presently limited by significant side reactions, poor quality deposition, and the potential to form hazardous dendrites. Therefore, it is important to develop a clear understanding of the surface reactivity and growth behavior of the lithium metal at the interface with the electrolyte in order to enable stable long-term cycling.

The products that form as a result of electrolyte decomposition reactions at the electrode interface are known to be extremely important in determining the final cell performance, yet characterizing the organic components of the solid electrolyte interphase (SEI) proves to be challenging with typical techniques. ATR-FTIR is a powerful method for characterizing organic species, but typical materials such as Ge, Si, and ZnSe react with lithium metal during deposition. In this talk, the development of a diamond based in situ ATR-FTIR cell will be discussed and its application to the characterization of organosulfur and fluorinated electrolytes will be discussed. Furthermore, SEI formation with respect to electrochemical potential and time will be described. With this understanding we provide new insights into the formation and chemical nature of SEI components that promote stable cycling of lithium metal electrodes.

450a Li Metal Solid State Battery Safety Assessed with Calorimetry and Modeling of Large-Format Cells

Paul Albertus

It is commonly assumed that the introduction of solid-state electrolytes implies the improved safety of high-energy batteries due to the removal of the combustible organic solvents in present Li-ion cells. However, safety is a multifaceted issue that involves thermal runaway onset temperature, self-heating rates, total heat production, maximum temperature, gas or other material release from a cell, thermal runaway prevention mechanisms, and other factors.[1-3] In addition, the use of lithium metal anodes (which melt at 180°C) for high energy density cells may introduce previously unconsidered safety issues due to the presence of molten and reactive lithium metal during a thermal runaway.

This talk will present results on the modeling of thermal runaway in a Li metal solid state battery, under both thermal ramp (e.g., in an oven) and short circuit conditions, with quantitative assessments of the mole fractions of key species in the cell and the temperature.[3] Key assumptions and parameter sensitivities will be shared, to clarify the most important areas for attention and analysis regarding Li metal solid state battery safety. In addition, results from differential scanning calorimetry (DSC) measurements of mg-scale quantities of solid state battery material sets will be presented, which provide quantitative information on onset temperatures and heat flows of key reactions. These experimental results will focus on cells with Li metal, an LLZO-based solid electrode, and metal oxide cathodes.

Finally, perspectives will be given on the early-stage (e.g., for materials and single-layer coin and pouch cells) evaluation of safety for prospective high-energy cells, for which commercial-scale cells (e.g., multi-layer cells of at least several Ah) are not yet available.

<u>*References*</u>

1. Bates, A.M., et al., *Are solid-state batteries safer than lithium-ion batteries?* Joule, 2022. **6**(4): p. 742-755.
2. Inoue, T. and K. Mukai, *Are All-Solid-State Lithium-Ion Batteries Really Safe?–Verification by Differential Scanning Calorimetry with an All-Inclusive Microcell.* ACS Applied Materials & Interfaces, 2017. **9**(2): p. 1507-1515.
3. Johnson, N. and P. Albertus, *Modeling Thermal Behavior and Safety of Large Format All-Solid-State Lithium Metal Batteries under Thermal Ramp and Short Circuit Conditions.* Journal of The Electrochemical Society, 2022. **169**(6).

450b Rechargeable Aqueous Zinc-Ion Batteries and Supercapacitors

Xiaolei Wang

Rechargeable battery technologies, particularly lithium-ion batteries (LIBs) are playing a dominant role in various application including electric vehicles, portable electronics and medical tools, as highly efficient power sources. LIBs are usually regarded as clean energy technologies, since they realize the conversion between chemical energy and electric energy in a 'green' manner without greenhouse gas emission. On the one hand, many transition metal elements (e.g., Ni, Co, Mn) are utilized in electrode materials, which have significant environmental impact. Moreover, a variety of chemicals (e.g., precursors for electrode materials, organic solvents for electrolyte, etc.) are consumed during battery fabrication, which makes them not 'green' and exacerbates the environmental deterioration. On the other hand, the uneven distribution of lithium and relevant transition metals reserves globally leads to the high cost of LIBs, while the highly active nature of lithium and the use of volatile organic electrolyte brings huge safety issues. Both high cost and safety concerns make LIBs not only challenging in electric vehicle application as a power battery technology, but also impracticable at all for large-scale electric energy storage as storage battery technology.

In order for batteries to make real and essential contributions to sustainability, it is highly desired to develop next-generation alternative battery technologies to LIBs with much lower cost and higher reliability. Aqueous batteries and supercapacitors have garnered tremendous research interest due to high safety. Among candidates, zinc-ion based energy devices show great promise due to the high theoretical capacity of Zn metal (820 mAh g−1), proper potential (−0.76 V vs. standard hydrogen electrode), water compatibility, and high safety of Zn metal. Despite the advantages, Zn-ion based energy storage devices are faced with issues from dendritic growth and low Coulombic efficiency in the Zn anode, fast capacity decay at the cathode, and unsatisfactory performance at extreme conditions. In this talk, we utilize electrode and electrolyte engineering to develop rechargeable low-cost Zn-ion batteries and supercapacitors with excellence electrochemical performance not only suitable for large-scale energy storage but also with potential for electric vehicle applications. In addition, we also try to realize recycling and upcycling of batteries, particularly ZIBs and LIBs, to create close loop of essential materials towards circular economy. Research focus has been concentrated on the recycling of LIBs electrode (particularly cathode) by regenerating or upgrading the electrode materials. Moreover, upcycling primary batteries into rechargeable batteries has also been studied.

450c Towards Enabling High Voltage $LiNi_{0.5}Mn_{1.5}O_4$ Cathodes: Understanding Transition Metal Dissolution Mechanisms with Operando NMR Spectroscopy

Lauren Marbella

The high-voltage $LiNi_{0.5}Mn_{1.5}O_4$ (LNMO) spinel cathode material offers high energy density storage capabilities without the use of costly Co that is prevalent in other Li-ion battery chemistries (e.g., $LiNi_xMn_yCo_zO_2$ (NMC)). Unfortunately, LNMO-containing batteries suffer from poor cycling performance because of the intrinsically coupled processes of electrolyte oxidation and transition metal dissolution that occurs at high voltage. In this talk, I will discuss our efforts to use operando electron paramagnetic resonance (EPR) and nuclear magnetic resonance (NMR) spectroscopies to show that transition metal dissolution in LNMO is tightly coupled to HF formation (and thus, electrolyte oxidation reactions as detected with operando and in situ solution NMR), indicative of an acid-driven disproportionation reaction that occurs during delithiation (i.e., battery charging). Leveraging the temporal resolution (s–min) of magnetic resonance, we find that the LNMO particles accelerate the rate of $LiPF_6$ decomposition and subsequent Mn^{2+} dissolution, possibly due to the acidic nature of terminal Mn-OH groups. X-ray photoemission electron microscopy (XPEEM) provides surface-sensitive and localized X-ray absorption spectroscopy (XAS) measurements, in addition to X-ray photoelectron spectroscopy (XPS), that indicate disproportionation is enabled by surface reconstruction upon charging, which leads to surface Mn^{3+} sites on the LNMO particle surface that can disproportionate into $Mn^{2+}{}_{(dissolved)}$ and $Mn^{4+}{}_{(s)}$. During discharge of the battery, we observe high quantities of metal fluorides (in particular, MnF_2) in the cathode electrolyte interphase (CEI) on LNMO as well as the conductive carbon additives in the composite. Electronic conductivity measurements indicate that the MnF_2 decreases film conductivity by threefold compared to LiF, suggesting that this CEI component may impede both the ionic and electronic properties of the cathode. Ultimately, to prevent transition metal dissolution and the associated side reactions in spinel-type cathodes (particularly those that operate at high voltages like LNMO), the use of electrolytes and electrodes that offer improved anodic stability and prevent acid byproducts will likely be necessary.

450d Revealing the Unknown Dynamics inside Lithium-Metal Batteries

Fang Liu, Yi Cui

The increasing demand for next-generation energy storage systems necessitates the development of high-performance lithium batteries. Unfortunately, current Li anodes exhibit rapid capacity decay and a short cycle life, owing to the continuous generation of solid electrolyte interface and isolated Li (i-Li). The formation of i-Li during the nonuniform dissolution of Li dendrites leads to a substantial capacity loss in lithium batteries under most testing conditions. Because i-Li loses electrical connection with the current collector, it has been considered electrochemically inactive or 'dead' in batteries. Contradicting this commonly accepted presumption, here we show that i-Li is highly responsive to battery operations, owing to its dynamic polarization to the electric field in the electrolyte. Simultaneous Li deposition and dissolution occurs on two ends of the i-Li, leading to its spatial progression toward the cathode (anode) during charge (discharge). Revealed by our simulation results, the progression rate of i-Li is mainly affected by its length, orientation, and the applied current density. Moreover, we successfully demonstrate the recovery of i-Li in Cu–Li cells with >100% Coulombic efficiency and realize $LiNi_{0.5}Mn_{0.3}Co_{0.2}O_2$ (NMC)–Li full cells with extended cycle life.

450e Direct Visualization of Lithium Transport in Nanostructured Ionic Liquid Electrolytes

Ryan Cashen, Matthew Gebbie

Battery fires are a growing problem as most modern batteries are composed of large amounts of flammable organic electrolytes. One avenue to resolve this issue is replacing volatile organic liquids with ionic liquids, which are non-flammable and chemically stable. However, typical ionic liquids exhibit slow battery charging kinetics, which is thought to arise from high viscosity causing poor lithium transport. Next generation ionic liquids that facilitate selective lithium transport by decoupling ion mobility from bulk viscosity promise key advantages in battery safety and performance. Yet, poor understanding of lithium transport in ionic liquids hinders control over lithium transference, in large part due to a lack of methods that selectively probe lithium mobility. Here, I will discuss a new diffusion-front mapping method that quantifies lithium diffusion in ionic liquids using fluorophores that change color with lithium concentration. Notably, this approach yields diffusion coefficients of metal ions in ionic liquids as a continuous function of concentration, contrasting prior methods that sample one concentration at a time. Further, we can evaluate concentration regimes that are inaccessible with other methods. Remarkably this approach promises to reveal more diffusion coefficients in a few months than have been reported in the previous decade and enables systematic investigation of lithium transport mechanisms in ionic liquids to determine why bulk viscosity is a poor predictor of lithium mobility. I will conclude by showing how this method opens the door to exploring new classes of ionic liquids that can bridge the advantages of traditional liquid and solid electrolytes.

450f Quantifying Reaction and Transport Heterogeneities between Individual Particles in Li-Ion Cathodes

Yiyang Li

Li-ion battery electrodes comprise of micron-sized solid particles immersed in liquid electrolytes. Understanding diffusion rates in the solid and reaction rates at the solid-liquid interface is essential for predictive modeling and rational design of battery materials. Unfortunately, measuring these transport rates in individual micron-sized battery particles has been extremely challenging. In this work, we present two approaches of reaction and diffusion rates at the single particle length scale. Our first approach uses operando synchrotron X-ray imaging of $LiFePO_4$ particles in organic liquid electrolytes, which uncovered strong autocatalytic behavior upon lithium de-insertion but autoinhibitory behavior upon lithium insertion. Our second, more recent approach is to measure the electrochemical response of many individual polycrystalline $Li(Ni,Mn,Co)O_2$ (NMC) particles using the microelectrode array. By using potentiostatic intermittent titration and fitting it to a spherical reaction-diffusion model, we were able to measure both the reaction and diffusion time constants. To our surprise, we found that neither the diffusion time nor the reaction time depended on the size of the particle, in stark contrast to the prevailing model. Instead, due to intergranular cracking, the electrolyte appears to penetrate into the secondary particle; as a result, the diffusion length is much shorter than the diameter of the secondary particle and does not change with the secondary particle size. These results have substantial implications for battery modeling and design.

AUTHOR INDEX

AIChE
120 Wall Street, FL 23
New York, NY 10005-4020

ISBN 978-1-7138-9297-7

9 781713 892977